ULTIMATE
ROBOT

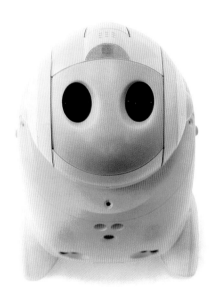

ULTIMATE
ROBOT

ROBERT MALONE

LONDON, NEW YORK, MUNICH,
MELBOURNE, DELHI

DK London
Project Editor Nicky Munro
Editor Chris Middleton
Designer Jenisa Patel
Managing Editor Adèle Hayward
Managing Art Editor Karen Self
Category Publisher Stephanie Jackson
Art Director Peter Luff
Production Controller Sarah Sherlock
Picture Researcher Carolyn Clerkin

DK Delhi
Project Editor Sheema Mookherjee
Designer Romi Chakraborty
DTP Coordinator Pankaj Sharma
DTP Designer Balwant Singh

First American Edition, 2004

Published in the United States by
DK Publishing, Inc., 375 Hudson Street,
New York, NY 10014

04 05 06 07 08 10 9 8 7 6 5 4 3 2 1

Copyright © 2004
Dorling Kindersley Limited

Text copyright © 2004 Robert Malone

A Cataloging-in-Publication record for this
book is available from the Library of Congress

ISBN 0-7566-0270-X

Reproduced by Colourscan, Singapore
Printed and bound in Hong Kong by Toppan

Page 1: PaPeRo; *Page 2*: Radicon; *Page 3*:

Tekno Cyber Dog Puppy; *Page 4*:
PaPeRos, Target Robot; *Page 5*: Sunday
Cruise, a painting by Eric Joyner

Discover more at
www.dk.com

Contents

Introduction

ROBOTS ARE HERE TO STAY. Once they were the fictional clanking mechanical men of early films, or the colorful toys that marched with sparking eyes and spinning gears. Today, robots are sleek humanoids with computers for brains—artificial beings that can walk, run, and talk like human beings. Even robot toys have become real robots that are small enough for children to play with. Digital technology is bringing us closer to realizing our futuristic fantasies. The robots that today's adults dreamed about as children are finally entering our workplaces and homes. They recognize our faces, and they speak our language.

Sico Millennia, entertainer

Packaging for a 1955 Japanese tin robot

Robots strike a precarious balance between humans and machines. They can be seen either as humanized machines, or as machinelike humans. To a large extent, robots are the result of humankind's desire to create beings in our own image: machines that look and behave as we do. Dramatic advances in computing and engineering mean that than those long-held ambitions are finally within reach.

While robots are obviously far less than human, they are much more than machines. Robots of all shapes and sizes are serious tools across a range of industries. They can build our cars, and protect our public spaces. Swimming and diving robots explore the ocean's depths, and microscopically accurate robot hands give

I, Robot, by Isaac Asimov, 1952

human doctors enhanced capabilities. Robots can fly over disaster zones and burrow through rubble. They can relay information to us about the farthest reaches of the solar system, and therefore be our eyes, ears, and hands on planets that are millions of miles away.

The boundaries between human and robot are being broken down every day. Digital technology has made it possible to create humanoid robots that can perform amazing feats. Robots can walk, run, talk, recognize faces, surf the Internet, climb stairs, play soccer, and get up if they fall down. They can even sulk, swear, dance, play games, and express emotions. Today's robots are designed to make us respond to them, and feel comfortable living with them. But it has not always been that way.

A reflection of fear

The word "robot" comes from the Czech word *robota*, meaning "forced labor." It was probably coined by Joseph Capek, playwright Karel Capek's brother, and was used in Karel's 1920 play *R.U.R.*—Rossum's Universal (or Automatic) Robots. The play is a futuristic nightmare in which machines and robots start replacing the working classes. Its success throughout the world seemed to reflect a deep-seated fear about the power of technology. Indeed, on stage,

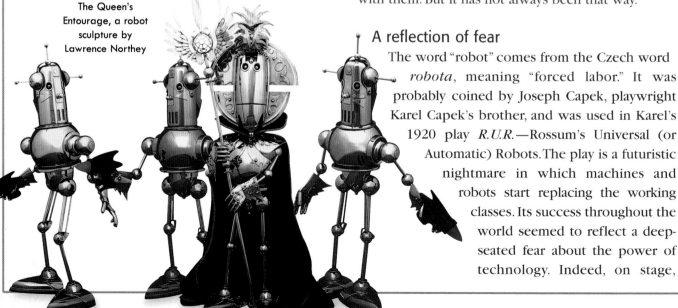

The Queen's Entourage, a robot sculpture by Lawrence Northey

Cybot, the DIY kit that learns as it grows

screen, and the printed page, robots have often represented what human beings might be like without emotions: soulless, dronelike machines—even remorseless killers.

A force for good

It took the writer Isaac Asimov to recognize that robots might be more than just the evil nemesis of mad scientists that populated movies and fiction, but also a potential force for good. In his short story *Liar!* (1941), he conceived the Three Laws of Robotics—a series of logical moral barriers that a computerized robot mind would have no choice but to respect. These fictional ideas were accepted by other writers and moviemakers, and are now seen as a blueprint for avoiding the problems of living with immortal, thinking machines. The laws are:

■ A robot may not injure a human being, or through inaction allow a human being to come to harm.
■ A robot must obey the orders given it by human beings, except where such orders would conflict with the First Law.
■ A robot must protect its existence, as long as this does not conflict with the First or Second Law.

Asimov later added a fourth law—the Zeroth Law—which states that a robot may not injure humanity, or through inaction, allow humanity to come to harm.

But whether they are evil or good, mechanical maniacs or saviors of the galaxy, rioting replicants, or hyperintelligent helpers, one thing is certain: robots continue to have a universal impact on the human imagination. Since the 1920s, robots have been movie stars and design icons. They have also been the subject of paintings, computer games, and sculptures.

Today, the robots are becoming closer and closer to their human creators, in both appearance and behavior. One thing is certain: robots can tell us a great deal about the times in which they are made. You might say that—ultimately—we get the robots we deserve.

Kawada Systems HRP-2 humanoid

Robert Malone

Robot snake

Myths and Machines

BEFORE ROBOTS COULD EXIST, centuries of innovation and cultural change had to occur. Several strands eventually combined to give robots "life": storytelling, the making of puppets and ritual masks, the invention of automation, and the desire to create "thinking" machines. These strands became increasingly intertwined during the Renaissance.

Mask of the Egyptian god Anubis

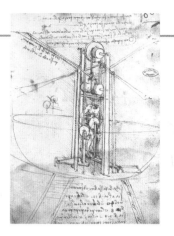

A machine with a pulley, designed by Leonardo da Vinci

For some, robots are superhuman machines; for others, they are monsters. Godlike stature and power have always fascinated mortals. Every culture has created its own myths that explore these ideas. Among these are the one-eyed Cyclops that Odysseus encounters in Homer's *Odyssey,* and Talos, the titanic, bronze statue that guards the island of Crete in several Greek myths. These, and later stories of robotlike statues, such as the Golem, a warrior made of clay, all have at their heart questions about whether humans can create life out of inanimate matter.

For centuries before robots appeared, many cultures evoked mythical figures using mechanical masks, puppets, and shadow figures. One of the earliest surviving examples is a mask of the Egyptian god Anubis, dating from 2,500 BCE. The mask had a movable jaw and a speaking tube in its

A design for a water-raising machine by Al Jazari of Persia, dating from 1206

mouth. Via the mask, Anubis could "appear" to his followers and "speak" to them, in reality through the human voice of a priest. In India and southeast Asia, the tradition of using mechanical puppets to "narrate" heroic tales to an awed populace dates back more than 2,000 years. Mechanical technologies and myths combined to endow the puppets themselves with great psychological power. That same power exists today in many fictional robots.

Early computers and calculators

Another step in robots' evolution was the the quest to make intelligent machines. The first stage was the invention of mechanisms that could log the movement of the stars, and therefore measure time. In the first-century, the Greek "computer" known as the Anticythera helped predict the position of the planets, the sun, and the moon by means of two rotating axles. Between 400 and 800 CE, the Islamic world perfected the astrolabe. This had many forms, the most common

A steam-driven door design by Hero of Alexandria

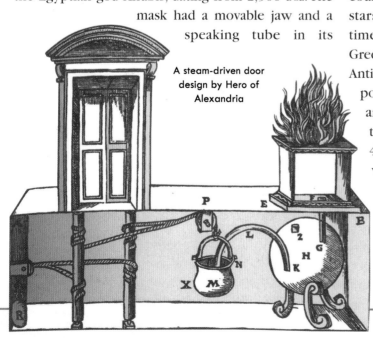

Articulated 19th-century wooden puppets from Indonesia

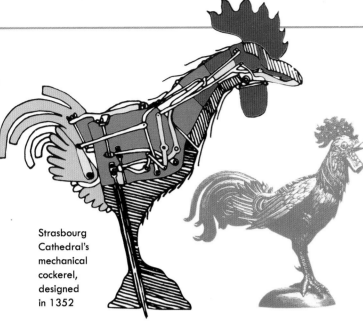

Strasbourg Cathedral's mechanical cockerel, designed in 1352

being a flat, mechanical arrangement of concentric discs that displayed the position of stars and planets at specific times. By the thirteenth century, engineers had designed clocks that measured time by means of interlocking gears. Many clockmakers also devoted their skills to building animated puppets, forging a link between accurate mechanisms and artificial beings. It was at this point that the separate strands of robots' ancestry began to merge. In 1352, an automated mechanical cockerel was added to the steeple of Strasbourg Cathedral. Every noon, it flapped its wings, opened its beak, stuck out its tongue, and crowed. This is the earliest surviving example of a time-telling, mechanical creature.

Additions to knowledge

In 1642, French scientist Blaise Pascal realized that the principle of using calibrated gearing to measure time could also be used to build machines that could display, and therefore add, numbers. He made the first mechanical adding machine, called the Pascaline. Its gears, ratchets, axles, drums, and external dials kept an accurate record of numbers inputted by a human operator. In 1674, a German mathematician, Gottfried

Leibnitz, unveiled the Stepped Reckoner—a simpler and more elegant solution to the same problem. Inventions such as these were the building blocks of programming and modern computer science, which is based on logging sequences of numbers and turning them into different types of data.

Automatic solutions

A further strand of robots' complex ancestry is automation—the science of building machines to carry out tasks with little or no human intervention. This is often thought of as a recent technology, but in fact, it is many centuries old. Two thousand years ago, Hero of Alexandria designed machines driven by steam, animal, or human power that could open doors, or drive a pipe organ. In 1206 in Persia, visionary engineer Al Jazari produced a book full of designs for machines that could lift water or serve food. Only a few were built, but the influence of all the designs was enormous.

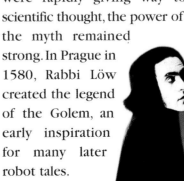

French philosopher and scientist Blaise Pascal

Creative engineering

Al Jazari was the most imaginative engineer the world had seen until the birth of Leonardo da Vinci in 1452. During the Renaissance—which championed human achievement—he designed numerous mechanical devices, including some automaton-style animals. He also sketched a mechanical man, and what is believed to have been a calculator—150 years before Pascal's machine. Although ancient beliefs were rapidly giving way to scientific thought, the power of the myth remained strong. In Prague in 1580, Rabbi Löw created the legend of the Golem, an early inspiration for many later robot tales.

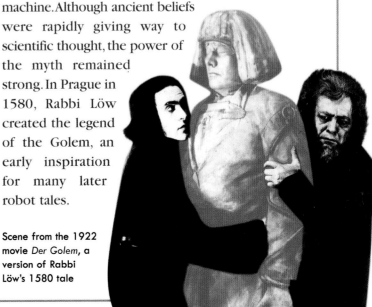

Scene from the 1922 movie Der Golem, a version of Rabbi Löw's 1580 tale

The Robot Revolution

ROBOTLIKE FIGURES APPEARED during the late Industrial Revolution, as people began to look for new ways of expressing the power of Victorian industry. But the first real and fictional robots were products of twentieth-century minds, when the revolution was one of computing and communications. Robots were the pinnacle of new technology, but for some were also a warning of its dangers.

The Steam Man by George Moore, engraved in 1882

The mid- to late eighteenth century was the golden age of the automaton. In France, Jacques de Vaucanson's Flute Player (1738), and Duck (1739)—which could walk, eat, and even excrete—were extremely sophisticated for the time, setting the standard for lifelike, animated figures driven by external machinery. Watchmakers Henri Jaquet Droz and his son Pierre went a stage further when they built internal mechanisms. Their automatons, the Lady Musician, the Draftsman, and the Writer (1750–1773), were delicate, self-contained, working figures. Automatons remained popular as entertainment figures for the rich in the nineteenth century, an era that was

Jacques de Vaucanson (1709–1782), with his automated duck

characterized by steam power, electricity, and industrial might. During this time, people became fascinated by technology's effect on their daily lives and jobs. Writers and artists joined engineers in envisioning the extraordinary future that was promised by new technology. Some of them believed that real humanoid machines would be the next logical step.

Visions and revolutions

Metal, electric, and automatic men began to march out of workshops and the pens of writers around the world. Some, like Louis Perew's huge walking automaton, Automatic Man (1900),

Robot Timeline

The history of robots is a story of science inspiring fiction, and fiction, in turn, spurring scientists on to make ever greater conceptual leaps. Automation, computing, and the Space Age were among science's contributions, while science fiction amazed us with its futuristic visions.

Popular culture

Tik-Tok and Dorothy

1907 The movie *The Mechanical Man and the Ingenious Servant* features a life-size mechanical Roman gladiator

1907 The second of L. Frank Baum's series of Oz stories is published, featuring Tik-Tok, the mechanical man

1910 The first movie of Mary Shelley's *Frankenstein* is directed by Thomas Edison

1912 The sculptor Brancusi creates *The Muse*, a stylized portrait of a face, which became a blueprint for Maria in *Metropolis* and other movie robots

Program announcing the release of Frankenstein

1920 Paul Wegener's movie *Der Golem* features a living statue of a warrior

1921 Karel Capek's play *R.U.R.*, written in 1920, is performed for the first time, in Prague

1926 Pioneering science fiction magazine *Amazing Stories* is launched

1926 Fritz Lang's *Metropolis* stars the female robot Maria

1900–1919

Technology

1900 The inventor Louis Perew unveils his Automatic Man automaton. It can walk and pull a carriage

1903 The first powered flight is made by Wilbur and Orville Wright

1913 Henry Ford installs the first moving conveyor belt-based assembly line in his car factory. A Model T can be assembled in 93 minutes

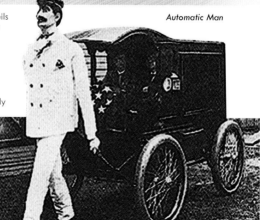

Automatic Man

1920–1929

1920 Public radio transmissions begin. By the end of the 1920s, 60 percent of families in the US have purchased a radio

1923 Vladimir Zworykin patents the iconoscope, an electric camera tube, starting the TV revolution

1927 Vannevar Bush invents the Differential Analyzer, the first mechanical/electric computer

actually existed, while others, such as George Moore's *Steam Man*, were fictional creations. At the turn of the twentieth century, children's book writer L. Frank Baum created the robotlike characters the Tin Man and Tik-Tok. However, it was only in 1920 that playwright Karel Capek coined a word to describe machines that resembled human beings: robot. His play *R.U.R.*, staged the following year, was a vision of society enslaved by technology. It was a timely work: powered flight and radio were both in their infancy, and television and computers were on the horizon. Artists and designers began developing the robot theme, seeing robots as icons of futuristic living. By the time the talking, smoking robot Electro appeared at the 1939 New York World's Fair, "robot" had become a household word.

Program for the first English production of Karel Capek's play *R.U.R.*, in 1923

technologically advanced worlds. In 1938, *Astounding Science Fiction* magazine published the first story to depict a sympathetic robot, *Helen O'Loy*. This inspired the young Isaac Asimov to write robot stories of his own. In these, he coined the word "robotics," and visualized how robots might one day coexist with humans.

Playthings and movie stars

Robot toys were hugely popular in the 1940s and 1950s. The first crude tinplate toys had appeared in the 1930s, and robot toys had quickly taken on a variety of forms that reflected the latest technological innovations. Robots had become icons. Soon, they became stars.

Buddy "L" Robotoy with its robot driver, the first robot toy

Robots of the imagination

During the late 1920s and 1930s, pulp magazines fed the public's appetite for robots. In 1929, publisher Hugo Gernsback coined the term "science fiction" to describe the new outpouring of stories about

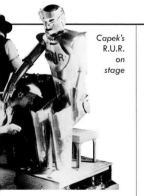

Capek's *R.U.R.* on stage

1930 The science-fiction magazine *Wonder Stories* is launched

1932 The Buddy "L" Robotoy is the first toy to depict a robot

1936 The Charlie Chaplin movie *Modern Times* warns of a future when machines will take over from humans in the workplace

1938 Lester del Rey's novel *Helen O'Loy* depicts a female robot as the ideal woman

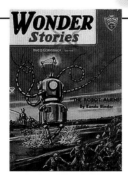

Wonder Stories magazine

1941 The movie *Man Made Monster* features a mad scientist who turns a man into an electronically controlled monster

1941 Isaac Asimov writes the short story *Liar!*, in which he lays down the Three Laws of Robotics

1949 In Jack Williamson's novel *The Humanoids*, humans rely so much on robots that their creativity and civilization is stifled. Williamson coins the word "humanoid"

Isaac Asimov

1930-1939

Elektro and the robot dog Sparky

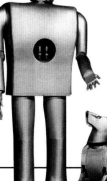

1923 radio

1932 Alpha the robot is unveiled at the Radio Exposition in London. With its internal radio, it seems to tell the time and "read" a newspaper

1937 Alan Turing's paper "On Computable Numbers" begins the computer revolution

1938 William Pollard and Harold Roselund invent a mechanical arm for an automated paint-spraying machine

1939 Elektro the robot is unveiled by Westinghouse at the New York World's Fair. He can smoke and talk

1940-1949

W. Grey Walter tortoise

1946 Robotics pioneer George Devol patents a general-purpose playback device for controlling machines

1947 John T. Parsons and Frank L. Stulen invent the numerical machine tool control (NC), the forerunner of today's computerized numerical controllers

1948 W. Grey Walter builds the first of his autonomous robot "tortoises." It can roam around, avoid obstacles, and react to light

The first generation of movie robots in the 1920s and 1930s were terrestrial creations, built by mad, human scientists who wanted to play God. By the 1950s, robots were increasingly portrayed as extraterrestrial, suggesting that people were beginning to feel alienated by technology and world events. In human history, this period encompassed the Wall Street Crash, World War II, the Atomic Age, the Cold War, and the rise of computing and the media.

Menacing machines

In cinema, the trend toward robots that represented fear of new technology continued, with movie robots increasingly portrayed as remote, alien figures. Robots seemed to point toward a dangerous future for humanity. People were uncomfortable with the idea of lifelike machines. Indeed, the more closely a robot resembled a human, the more sinister it was perceived as being. Machine-like robots were not so threatening.

QRIO demonstrates its ability
to hold objects and dance

Real robots arrive

Toward the end of the 1940s, scientists began turning fiction into reality by building the first intelligent machines. In 1946, industrialist George Devol patented a means of controlling machines with instructions stored on magnetic tape. This was the first step in robots' shared journey with mainframe computers, and, later, with PCs. In 1948, scientist W. Grey Walter's experiments in artificial life led to the development of the first autonomous robots. His robotic "tortoises" reacted to and followed light, appearing to exhibit intelligence. These were the forerunners of today's popular kit robots that respond to external stimuli. In 1961, Devol and Joseph Engelberger unveiled Unimate, the first industrial robot—a robotic arm designed to work on the assembly line at General

Popular culture

1950 Asimov's book *I Robot* is published

1951 Gort appears in the science fiction epic *The Day the Earth Stood Still*

1954 In the movie *Tobor the Great*, a benign space exploration robot is stolen by enemy agents

1956 Robby the Robot stars in *Forbidden Planet*

Movie poster, 1954

1961 *Solaris*, a novel by Stanislaw Lem, features human replication

1966 Swiss artist H.R. Giger produces "biomechanoid" art, depicting beings that are part human, part machine

1968 *2001: A Space Odyssey*, a movie by Stanley Kubrick, introduces the malign computer HAL

1968 The story *Do Androids Dream of Electric Sheep?*, by Philip K. Dick, features replica humans. It becomes the basis for the movie *Bladerunner* in 1982

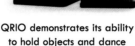

1972 *The Stepford Wives*, a novel by Ira Levin, features women who are replaced by robots. It is filmed in 1975, and again in 2004

1976 Isaac Asimov's novel *Bicentennial Man* features a robot that becomes human

1977 The first of the *Star Wars* movies is released

R2-D2

1950-1959

Technology

1951 Raymond Goertz designs the first "tele-operator," an articulated arm

1954 George Devol designs the first programmable robot

1956 Devol and Joseph Engelberger set up the first robot company, Unimation

1957 Sputnik I, the world's first artificial orbiting satellite, is launched by the Soviet Union, starting the space race

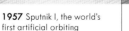

Sputnik before launch

1960-1969

1962 Unimate, the first industrial robot, joins the assembly line at General Motors

1964 The IBM 360 is the first computer to be mass-produced

1968 Unimation licenses its technology to Kawasaki, starting an explosion of robot development in Japan

1969 Neil Armstrong becomes the first person to walk on the Moon

Neil Armstrong

1970-1979

1970 The Stanford Research Institute's (SRI's) Cart and Shakey are the first computer-controlled robots

1974 Vicarm markets a microcomputer-controlled arm for industrial applications

1975 The Viking I Mars orbiter and lander is launched.

The launch of Viking I

Motors. Within a few years, Japanese and American robots could weld, assemble, and spray complete cars without human intervention. The key to this was Artificial Intelligence.

Robots with brains

Artificial Intelligence (A.I.) as a discipline was first conceived in 1957, and is the foundation of modern robotics. A.I. allows robots to follow a sequence of preset actions independently. In real robotics, this is the result of programming, rather than self-determination, but movie-makers were quick to explore the threat from machines that "think" for themselves. An early example is HAL, in *2001: A Space Odyssey* (1969). HAL—an evil, autonomous computer—attempts to kill all the astronauts on a spaceship. It was a chilling vision, but HAL's real contemporaries, the first PCs, lacked the processing power to operate even a robotic arm

Computer artwork showing the future NASA Skyworker robots constructing a space facility in near Earth orbit

successfully. Today, however, there are many robots working in space, and NASA is designing Skyworker robots to build and repair orbiting space stations.

In 1965, George E. Moore, co-founder of microchip manufacturer Intel, said that the number of components that could be put onto a silicon chip would double every 12 months (he revised that to 18 months in 1975). "Moore's Law," as it became known, explains why micro-processors have become so fast and powerful in such a short time. During the 1980s, the link between robotics and computers was reinforced with the first programmable "personal" robots. Today's fast processors, digital systems, and the Internet mean that robots have not just genuine intelligence, but also access to mobile communications. A generation of walking, talking, humanoid robots, such as Honda's ASIMO and Sony's QRIO, finally links robots' real and fictional selves.

1982 The movie *Bladerunner* features android "replicants"

1983 *Tik-Tok*, a novel by John Sladek, features an artistic robot awaiting trial for murder

1984 The first *Terminator* movie is released

1984 GoBot toys debut

GoBot Dumper

Sony AIBO

1997 The first RoboCup football tournament is held in Japan

1998 LEGO launches its first Robotics Invention System

1999 The first of Sony's AIBO robotic dogs goes on the market

2001 The movie *A.I.* features a robot boy who longs to be human

2003 Wow Wee introduces RoboSapien, a humanoid toy

2004 ZMP Inc. releases Nuvo, a robot with a cell-phone link

Nuvo

Popular culture

1980-1989

1983 Small mobile robots for home and educational use are made in the U.S.

1984 SRI's Flakey reflects the dramatic improvements in state-of-the-art technology

1986 Honda launches a project to build a walking humanoid robot

Flakey

1990-1999

1990 Unmanned Air Vehicles (UAVs) are developed by the US and Israel

1994 Dante II, an eight-legged walking robot, descends into the crater of Mt. Spur volcano

1996 Honda's P2 humanoid, a prototype for ASIMO, debuts

Dante II

2000-present

2000 NASA unveils Robonaut, a remote-controlled "space helper"

2002 A flying robot for Mars exploration is developed by Anthony Colozza

2004 The DARPA Grand Challenge, a competition for autonomous vehicles, takes place. No team completes the course

Robonaut

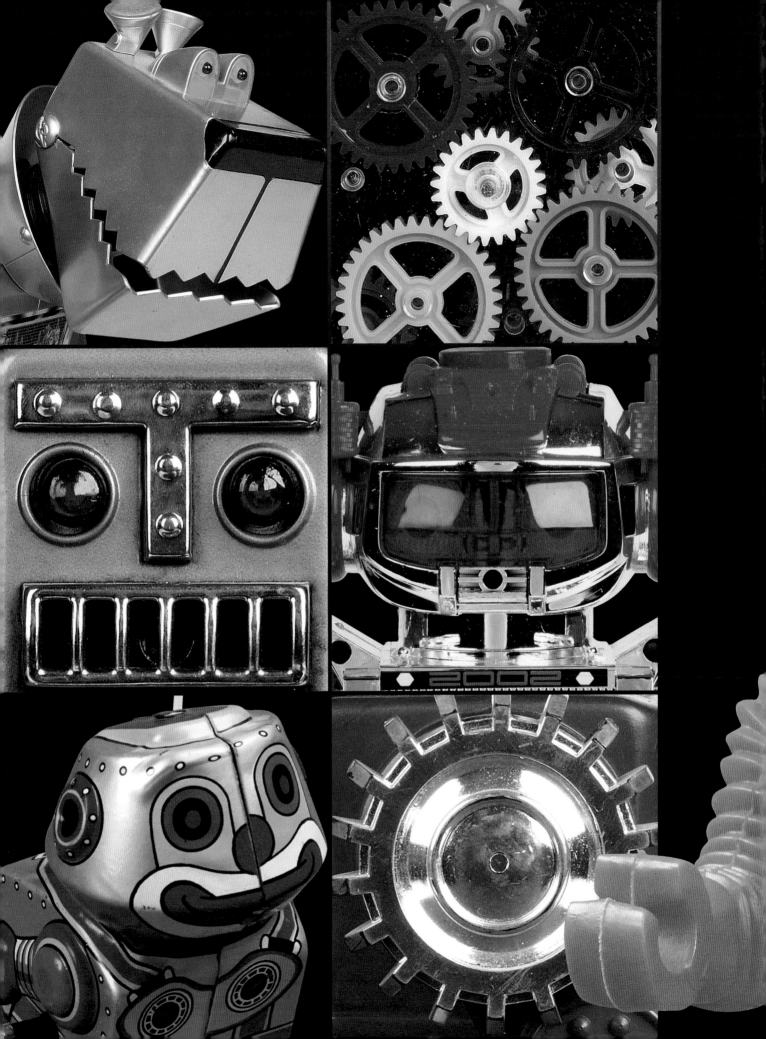

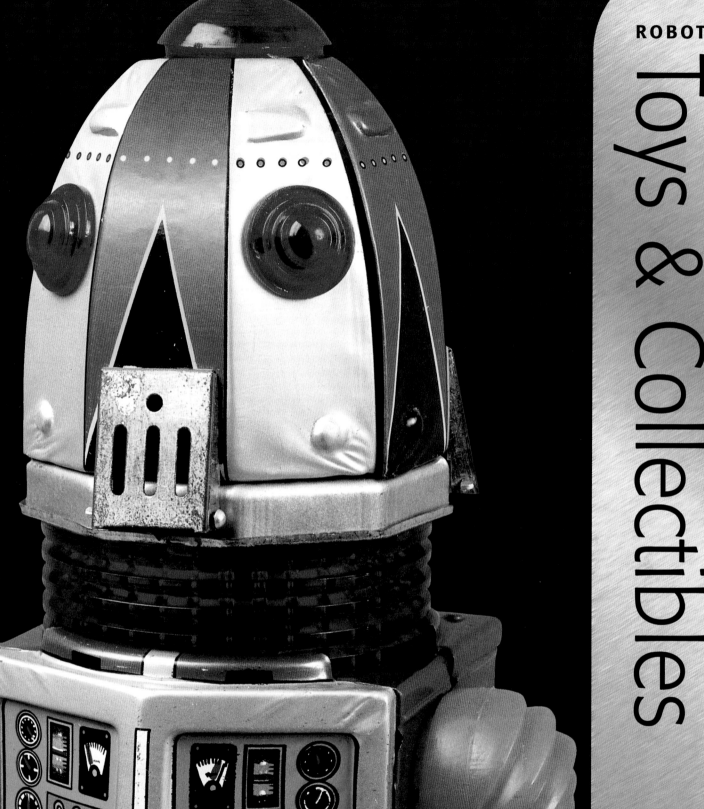

Toys & Collectibles

Tinplate to Techno Toys

THE FIRST ROBOT TOY was the Buddy "L" Robotoy (1932), a tinplate truck with a robot driver. Soon afterward, a host of clockwork robots and then battery-driven robots marched into children's lives. In the decades since, robot toys have continued to amuse, charm, and thrill each generation. Many of today's toy robots can walk, talk, and respond to commands. Some are effectively real robots in miniature.

A 1950s remote-controlled toy

The idea of robots being "metal men" was popularized by the 1920 play *R.U.R.*, which had a huge impact on the public's imagination. Although *R.U.R.*'s humanoid machines created a prototype for robot toy design—half man, half machine—there was no precise blueprint. As a result, designers' imaginations could reign supreme.

The first "pure" robot toy was probably the colorful clockwork Robot Lilliput, first produced in the mid-1930s. This was followed swiftly by the warlike Atomic Man from Japan, the smiling Nando from Italy, and a legion of other tin toys.

Invincible warriors

Lithographed tinplate toys were cheap to produce during World War II and the economically harsh years afterward. Today, many such toys are poignant echoes of the times in which they were made. In many ways, the first wave of robot toys resembled warriors and war machines, with armor-like bodies and industrial details. Their simple mechanisms made them move stiffly and slowly—a feature that influenced the way people thought about robots for decades afterward.

But many early toy robots could do more than just walk until their springs unwound. Some sparked,

others had spinning gears or pumping pistons. With the introduction of battery power in the late 1940s and remote control in the early 1950s, some robots could light up, revolve, fire ray guns with sound effects, and even emit smoke from their mouths. As plastic and, later, die-cast metal became popular materials, robot toy design became even more colorful, intricate, and detailed.

A 1970s wind-up plastic robot

Big Loo, a 3-ft- (1-m-) tall robot toy from 1965

Merchandising opportunities

The link between robot toys and their counterparts on the big and small screens began with Robby, star of the 1954 movie *Forbidden Planet*. Robby was the first robot celebrity, and a host of Robby and Robby-like toys appeared as manufacturers cashed in on his popularity. Once the concept of a series of related toys had been established by the various Robby spin-offs, manufacturers seized on the idea of producing groups of related toys that children would want to collect. Soon, "gangs" of robot characters began to appear, beginning with the tinplate Gang of Five. This trend continued through the 1960s and 1970s, with gangs such as the plastic, battery-operated

A late-1970s Golden Gear Robot

Zeroids. The 1980s saw the arrival of transforming robots, including GoDaiKin, Transformers, and GoBots, that were designed to switch from a robot to an entirely different entity—two toys in one. These toys reforged the link between robots and warriors or war machines that had originally been made by the first robot toys. Bandai's GoDaiKin robots bear a strong resemblance to Shogun and Samurai fighting men, and many transforming toys changed into tanks, fighter planes, or armored trucks. Each of the robots was marketed as being a character in a wider story that encompassed heroes and villains and battles and adventures. Films and TV series based on them further increased their popularity.

The film/toy merchandising tie-in is advantageous to both manufacturer and studio, with movies fueling the toys' popularity and vice versa. The most successful is the *Star Wars* films and their famous droids.

Super GoBot, Bug Bite

Reawakening the past
In recent years, there has been renewed interest in the toys of the past. While digital technology has created a generation of "smart" computerized toys, many people still admire the simpler charms of tinplate robots. Alongside the mass of advanced robot toys, there is also a thriving market for reproduction tin toys. With prices soaring, many original tin toys are far beyond the reach of most people's budgets. Reproduction toys allow people to enjoy the nostalgic appeal of tinplate robots, but at toy, rather than antique, prices.

C-3PO and R2-D2 collectible figures

Real robots as toys
Now that cheap microprocessors and miniaturized electronics are commonplace, many of today's robot toys are sophisticated machines packed with computer technology. This new generation of toys is growing into a family of real robots in miniature. Many use sensors, digital cameras, and voice-recognition technology to interact with their owners, and some can even develop "personalities" as they learn about their environments. Sony's AIBO dog was the first such robot, but today there are many other toys that walk, talk, react to light and sound, recognize voices, and even connect to the Internet. The next generation of robots will further blur the distinction between toys, entertainment devices, and robot companions. So, while many adults are seeing their childhood fantasies come true, for today's children, these high-tech toys are everyday playthings that point to an exciting future of their own.

A Sony AIBO robot dog

Lavender Robot, one of the Gang of Five tinplate toys from Japan

Robot Lilliput

The Japanese tinplate Robot Lilliput was probably the first true robot toy—and was certainly the first to be mass-produced. Dating from the mid- to late 1930s, Lilliput has a bright color scheme and a friendly, almost human face, unlike many of the more austere-looking toy robots produced during the war years.

The pioneer

High-quality for the time it was made, this simple wind-up toy marks the start of a golden age of toy robots. Lilliput's legs are rigid, but on the base of each foot is a short prong. When the spring is wound, the prong extends and retracts, first in one foot, then the other, propelling the robot forward. The awkward jerkiness of its movement is typical of early robot toys.

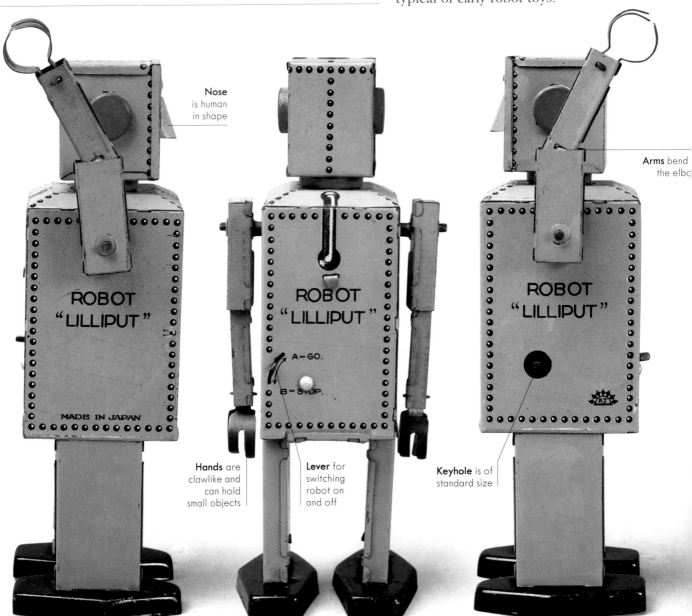

Nose
is human
in shape

ROBOT
"LILLIPUT"

MADE IN JAPAN

ROBOT
"LILLIPUT"

A = GO.

B = STOP.

Hands are
clawlike and
can hold
small objects

Lever for
switching
robot on
and off

Arms bend
the elbo

ROBOT
"LILLIPUT"

Keyhole is of
standard size

▲ **Parallel lines**
Viewed in profile, Robot Lilliput's simple, functional design is revealed as a series of different box shapes. The effect is stocky and machinelike.

▲ **Revealing back**
The on/off switch is on Lilliput's back. Also on the back is an unpainted tab, showing that some parts were painted before the toy was assembled.

▲ **Common keys**
Since keys of wind-up toys were easily lost, they usually came in standard sizes that fitted a variety of toys. Children built up a collection of keys as well as toys.

Head is box-shaped

Eyes are set in a square shape for a robotic look

Teeth are bared but not alarming

Ears are simple discs

Arms can be rotated manually

Chest detailing and serial number are identical on all models

Feet are a distinctive hexagonal shape

N.P. 5357.

Specification: Robot Lilliput

First manufactured:	1930s
Country of origin:	Japan
Manufacturer:	Unknown
Height:	6 in (15 cm)
Power source:	Wind-up
Features:	Pin-driven walking action

◀ Painted features

One of the most collectible of the tin robots, Lilliput has a simple color scheme of black trim on an orange-yellow body, legs, arms, and head. Red ears, neck, and mouth add a cheerful contrast. The chest has lithographed dials, rivets, a serial number, and even an air hose. The claw-like hands allow it to hold objects such as pencils.

Mouth differs from the original robot

N.P. 5357.

▲ Collector's item

In 2001, Tin Tom Toys of China brought out a reproduction aimed at adult collectors. It is high-quality, but not an exact copy, being slightly broader and lighter in color than the original.

Protruding ears are unique among robot toys

Helmet has a distinctive military look

Arms can be rotated manually

Lithographed details add color

Stop/Go lever, printed in English, activates the toy

Broad feet aid balance

Early Tin Toys

Many manufacturers of tin robot toys sprang up in post-war Japan and Europe. They used lithographed tinplate, which allowed for simple shapes and surface decoration, and was cheap to produce on a small scale. These first tinplate robots were humanoid in shape and often military in appearance, reflecting the era in which they were made.

Atomic Man

Now comparatively rare, this Japanese toy was constructed of pressed tin, and tabbed together, piece by piece. Atomic Man has an internal clockwork mechanism that allows it to walk in a way similar to early Japanese walking dolls, which were driven by bamboo gears. Lithographed details on the chest and printed rivets enliven the otherwise austere design, which is reminiscent of the steam age and of early-twentieth-century war machines.

◄ **Modified movement**

As well as a walking mechanism, Atomic Man has asymmetrically designed arms—one arm is crooked—that can be rotated manually. The chest is printed with dials and gauges and a clock set forever at 1:22.

Specification: Atomic Man	
First manufactured: 1940s	
Country of origin: Japan	
Manufacturer: Unknown	
Height: 5¼ in (13 cm)	
Power source: Wind-up spring action	
Features: Walks, arms can be rotated	

Nando

This is the only known Italian robot toy produced in the years immediately after World War II. Nando has an unusual means of control: pneumatic propulsion. The robot walks and turns its head, but the arms are fixed, crooked at the elbow. Few Nandos have survived intact, especially if they have been heavily used. Cheaply produced at the time, this robot is today highly collectible, more so if the rubber bulb and the box are in good condition.

▶ **Lasting appeal**

The boxy design makes use of bare metal, with the riveted chest dominated by a grille, and the rectilinear legs ending in huge feet. Eye, nose, and mouth decals (transfers) create a friendly face above the rugged bodywork.

Rubber bulb activates the leg and head movement

▲ **Pneumatic propulsion**

When squeezed, Nando's rubber bulb drives air down the tube into the mechanism, moving the robot's legs and turning its head. This design is used in many plastic toys, but its use in a robot is rare.

Antenna is made from a simple wire hook

Facial decals add color and appeal

Breastplate in bare metal looks weathered, but is part of Nando's charm

Hands echo box-like construction of the body

Grille on the chest imitates industrial machinery

Rivets are pressed from tin

Blocklike foot design was much imitated by later manufacturers

Specification: Nando

First manufactured:	Early 1950s
Country of origin:	Italy
Manufacturer:	OPSET
Height:	5¼ in (13 cm)
Power source:	Pneumatic propulsion, with air pumped from handheld bulb
Features:	Walks, turns head

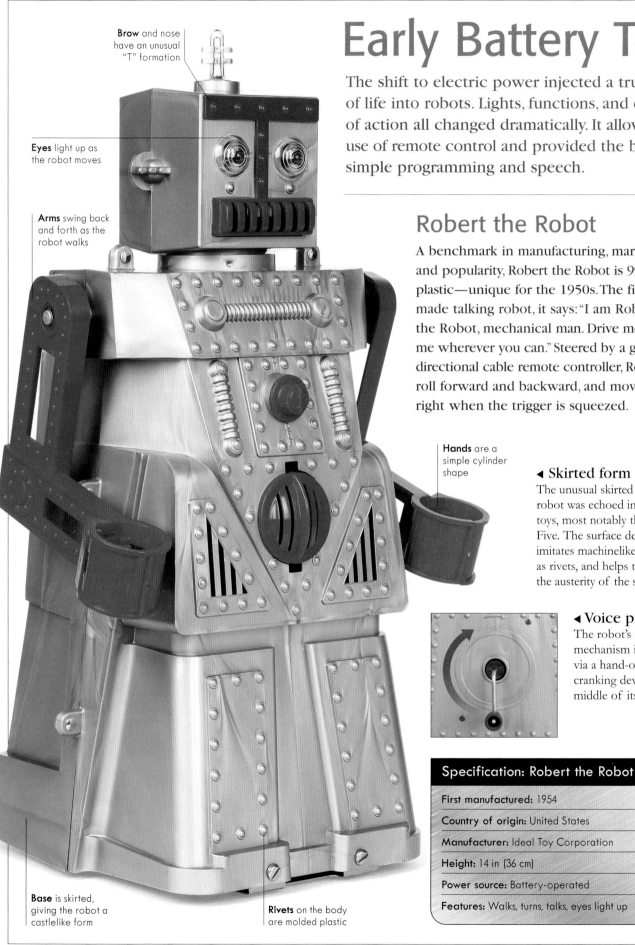

Brow and nose have an unusual "T" formation

Eyes light up as the robot moves

Arms swing back and forth as the robot walks

Hands are a simple cylinder shape

Base is skirted, giving the robot a castlelike form

Rivets on the body are molded plastic

Early Battery Toys

The shift to electric power injected a true spark of life into robots. Lights, functions, and duration of action all changed dramatically. It allowed the use of remote control and provided the basis for simple programming and speech.

Robert the Robot

A benchmark in manufacturing, marketing, and popularity, Robert the Robot is 99 percent plastic—unique for the 1950s. The first US-made talking robot, it says: "I am Robert the Robot, mechanical man. Drive me, steer me wherever you can." Steered by a gun-shaped directional cable remote controller, Robert can roll forward and backward, and move left and right when the trigger is squeezed.

◄ Skirted form
The unusual skirted form of this robot was echoed in many later toys, most notably the Gang of Five. The surface decoration imitates machinelike detail such as rivets, and helps to lighten the austerity of the shape.

◄ Voice program
The robot's talking mechanism is activated via a hand-operated cranking device in the middle of its back.

Specification: Robert the Robot

First manufactured:	1954
Country of origin:	United States
Manufacturer:	Ideal Toy Corporation
Height:	14 in (36 cm)
Power source:	Battery-operated
Features:	Walks, turns, talks, eyes light up

Zoomer

This tinplate toy with the bearing of a military machine and the tools of a mechanic harks back to the simplicity of Atomic Man and Lilliput. Its form and detail are similar to its contemporaries ST-1 of Germany, and Sparky. Zoomer walks forward clutching a wrench, and its plastic eyes light up. On its back is a pressed tin backpack with an extendable antenna.

▶ **Minimalist styling**
Zoomer's stark appearance is typical of an early robot, with plain tinplate bodywork, gaping eyes and mouth, and arms pointing forward like a sleepwalker's. The faceplate is attached like a mask, adding to the distinctive, impersonal expression.

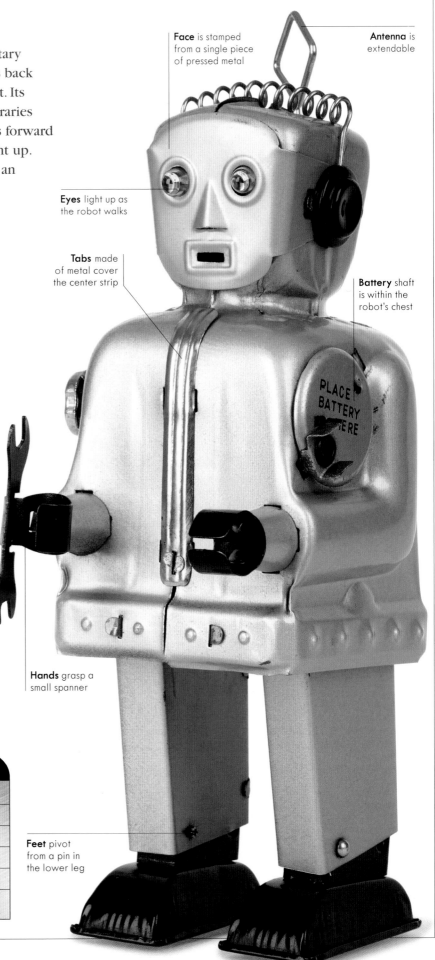

Face is stamped from a single piece of pressed metal

Antenna is extendable

Eyes light up as the robot walks

Tabs made of metal cover the center strip

Battery shaft is within the robot's chest

PLACE BATTERY HERE

Hands grasp a small spanner

Feet pivot from a pin in the lower leg

▲ **Rugged setting**
The box shows Zoomer walking through a devastated city with a hammer in its hand and eyes beaming. Alternative packaging shows the robot in an icy landscape with shafts of lightning.

Specification: Zoomer

First manufactured:	1950s
Country of origin:	Japan
Manufacturer:	Nomura
Height:	8 in (20 cm)
Power source:	Battery-operated
Features:	Walks, eyes light up

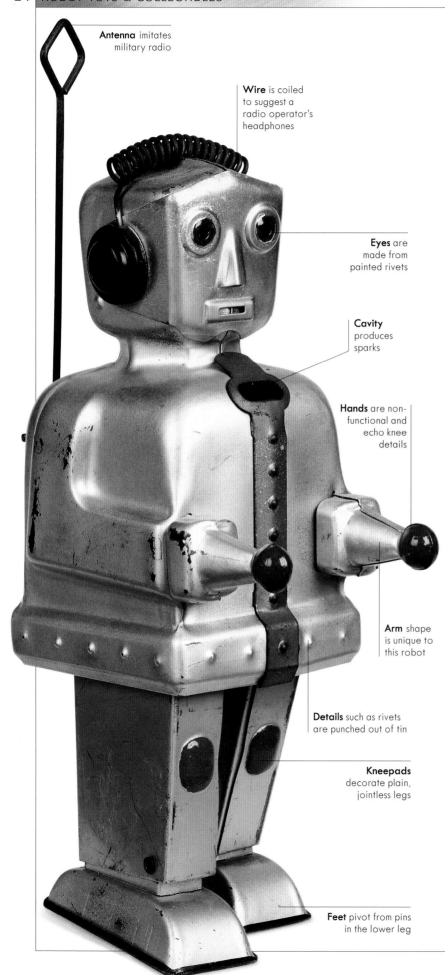

Antenna imitates
military radio

Wire is coiled
to suggest a
radio operator's
headphones

Eyes are
made from
painted rivets

Cavity
produces
sparks

Hands are non-
functional and
echo knee
details

Arm shape
is unique to
this robot

Details such as rivets
are punched out of tin

Kneepads
decorate plain,
jointless legs

Feet pivot from pins
in the lower leg

Sparking Robots

Clockwork sparking toys have been common since the 1950s, but sparking robots are comparatively rare. The wind-up mechanism drives a spinning flint that creates a sparkler effect as the robot moves. Toymakers Yoshiya in Japan and Strenco in Germany used this simple technology to bring their austere tin men to life.

Robot ST-1

This rare 1950s German toy shares certain characteristics with other robots that were being produced in Japan and elsewhere at the time. The masklike face and the style of the head, body, antenna, and headphones were all popular with contemporary toy manufacturers. To make it inexpensive to produce in what was a harsh economy, Robot ST-1 is largely unpainted, but there are flashes of color and originality. Among these are the red, knob–shaped hands, reminiscent of a Van de Graaff electrostatic generator—or angry fists.

◄ **Spring driven**
When it is wound up, Robot ST-1's spring drives an axle and gears as it unwinds, moving the legs back and forth by means of rods attached by pins. It uses a small fiber flap in its foot to move itself forward.

Specification: Robot ST-1	
First manufactured:	1950s
Country of origin:	Germany
Manufacturer:	Strenco
Height:	8 in (20 cm)
Power source:	Clockwork
Features:	Walks, sparks

Sparky

Both Sparky and Robot ST-1 are more machine-like in appearance compared with other toy robots of the same period, which were made to appear more human, with features such as knees and bending elbows. Like Robot ST-1, Sparky is unpainted, and the touches of color are in its bright red eyes, feet, and headphones.

▶ Minimalist design

This largely unpainted clockwork toy is basic but attractive. Made of tinplate, it has simple ringed hands, and a masklike face held on by headphones. It emits sparks from its eyeholes as it walks, and the action stops when the antenna is pushed down.

Shoulder joints are rivetlike

Arms can be moved manually

▲ Laser vision

Sparky's packaging artwork shows its eyes blazing more fiercely than the toy's eyes are actually capable of doing. Once out of the box, the robot appears rather more benign than the packaging suggests.

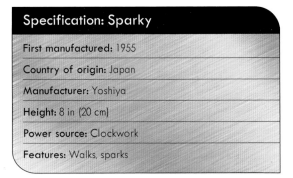

Specification: Sparky

First manufactured:	1955
Country of origin:	Japan
Manufacturer:	Yoshiya
Height:	8 in (20 cm)
Power source:	Clockwork
Features:	Walks, sparks

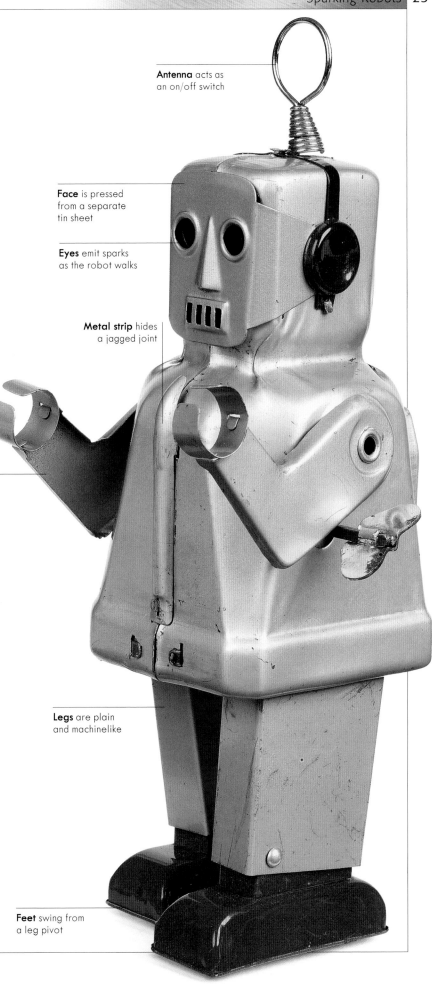

Antenna acts as an on/off switch

Face is pressed from a separate tin sheet

Eyes emit sparks as the robot walks

Metal strip hides a jagged joint

Legs are plain and machinelike

Feet swing from a leg pivot

The Gang of Five

These five large robots, produced by the same manufacturer, are all essentially of the same design with variations in color and detailing. Each has a different function that gives it character. They were thought to be a Gang of Four, until the previously lost Machine Man was rediscovered in the 1990s.

The gang members

These toys are decorated with high-quality tinplate offset lithography. Each robot has a different feature. Radicon is radio-controlled; Target Robot changes direction after its target has been hit, Giant Sonic Robot has a loud train whistle, and Machine Man has a bump-and-go action, as does Nonstop Robot, which is also known as Lavender Robot.

▼ Unity in diversity

The robots have a skirted design. Machine Man and Nonstop Robot display Cubist designs. In contrast, Radicon has an austere gray form. The violet Target Robot has a red circular target on its chest for shooting at. Giant Sonic Robot, finished in red with black head and arms, has a much larger head than the others.

▲ Radicon's radio control

Radicon is the first completely radio remote-controlled toy, although its simple electronics are primitive by modern standards and can be rather unreliable. A battery-operated remote controller communicates with the robot, sending commands such as Forward, Stop, Left, and Right.

Machine Man

Radicon

Giant Sonic Robot (Train Robot)

Target Robot

Specification: Radicon

First manufactured: 1956

Country of origin: Japan

Manufacturer: Matsudaya

Height: 15 in (38 cm)

Power source: Battery-operated

Features: Moves by remote control

▶ Collectible robot

This hammered gray, industrial-looking robot is by far the least decorative of the Gang of Five, but its remote control capabilities make Radicon very desirable to collectors. The large ear antenna recalls the design of Robby the Robot, who appeared in the same year. Radicon's arm construction is more complex than those of other gang members, since it has an elbow panel holding the upper and lower arm together, allowing a greater range of movement.

Eyes and other facial features add color

Panel in chest has an on/off switch and a working dial

Antenna receives a signal from the remote control

Upper arms swing as the robot moves

Elbow joint allows for limited manual lower arm movement

OFF – ON

Batteries are housed beneath a removable screw-on panel

Nonstop (Lavender) Robot

TV Screen Robots

In the 1950s, television was an exciting new global phenomenon, and toy manufacturers were quick to exploit its popular appeal. Neither technology nor economics allowed for real TV's to be built into toys, so the alternative was to use imitation screens to create the illusion of images broadcast from deep space and alien worlds.

Space Explorer

In 1954, color television broadcasts began in the US, and Russia launched Earth's first artificial satellite, *Sputnik*. Space Explorer appeared between this landmark year and *Sputnik's* first broadcasts in 1957, forging a connection between fantasy and new media, just as the first robot toys had done with radio.

▶ First transforming toy

Space Explorer was the first robot to transform completely into a different object, five years before the automatic transformer Space Robot (Tulip Head), and nearly 30 years before the Transformers' and GoBots' hand-manipulated identity changes. The toy's feet are locked together to form the TV base, and it moves in a pin-driven, waddling motion.

Specification: Space Explorer

First manufactured:	1955
Country of origin:	Japan
Manufacturer:	Yonezawa
Height:	11¼ in (28.8 cm)
Power source:	Battery-operated
Features:	Walks, displays illuminated screen, transforms from robot to "TV set"

Head disappears inside in TV mode

▶ TV mode

When Space Explorer is put into TV mode, its arms retract, its body lowers on to its feet, and its head drops into the body, leaving only a small section of it exposed.

Head design is unique to this toy

Eyes flash when the robot is switched on

Arms swing up in robot mode

Screen is illuminated by a bulb showing a 3-D effect scene

Body, arms, and legs are all-steel

Feet include waddling mechanism

Television Spaceman

This 1961 robot is a mechanical marvel, with its moving TV pictures achieved by a backlit paper loop, and gear-driven eyes that spin in opposite directions. It is an excellent example of how plastic can be used to create intricate details over a metal framework. With its eye-catching antenna, this robot is an extravagant mix of space-age style and period design.

▶ Heading for the moon

This robot's screen shows a space scene with a rocket positioned for launch—eight years before humans set foot on the Moon. The pylonlike antenna suggests power, but seems a fragile addition to such a heavy-duty toy. An attractive touch is the fearsome serrated mouth, which is lit from within by chaser lights as the robot strides out on its galactic mission.

Antenna top is fragile and easily damaged

Eyes have spinning gears

Antenna doubles as an on/off switch

Screen acts as a magnifying lens for the images within

Arms swing as the robot walks

Hands are nonfunctional design feature

Dials are painted on and nonfunctional

Hands can be moved manually

Feet are hinged and conceal the inner wheels

Specification: Television Spaceman

First manufactured: 1961

Country of origin: Japan

Manufacturer: Alps

Height: 14¾ in (36.8 cm)

Power source: Battery-operated

Features: Walks, swings arms, spins eyes, displays moving pictures on TV screen, has illuminated mouth

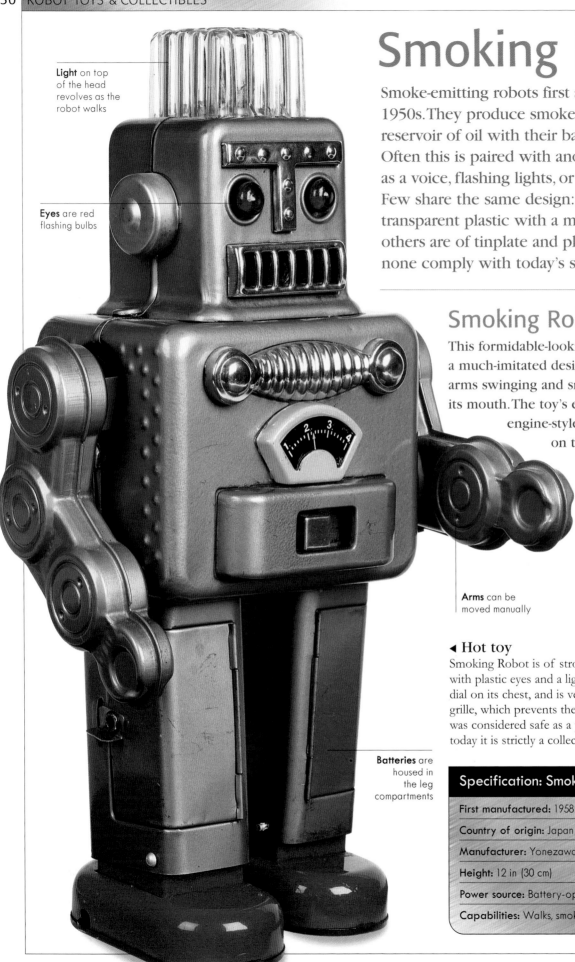

Light on top of the head revolves as the robot walks

Eyes are red flashing bulbs

Arms can be moved manually

Batteries are housed in the leg compartments

Smoking Robots

Smoke-emitting robots first appeared in the 1950s. They produce smoke by heating a small reservoir of oil with their batteries' low voltage. Often this is paired with another feature, such as a voice, flashing lights, or a walking action. Few share the same design: some are of transparent plastic with a metal skeleton; others are of tinplate and plastic. Of course, none comply with today's safety standards.

Smoking Robot

This formidable-looking smoking robot has a much-imitated design. It walks forward, arms swinging and smoke streaming from its mouth. The toy's eyes flash and a fire-engine-style warning light spins on top of its head. Many toys since have copied the basic boxy, vehicle-style design, usually without the smoking function or other attractive features.

◀ Hot toy
Smoking Robot is of strong tinplate construction, with plastic eyes and a light. It sports a nonfunctional dial on its chest, and is ventilated by a bumperlike grille, which prevents the toy from overheating. It was considered safe as a plaything at the time, but today it is strictly a collectors' item.

Specification: Smoking Robot	
First manufactured:	1958
Country of origin:	Japan
Manufacturer:	Yonezawa
Height:	12 in (30 cm)
Power source:	Battery-operated
Capabilities:	Walks, smokes, light spins and flashes

Chief Smoky

Instead of putting out fires, this 1960s fire-fighter robot generates smoke from the top of its helmet. It has a bump-and-go action, changing direction whenever it hits an obstacle. The toy is mainly made of tinplate, with the exception of the eyes and head panels, which are of clear plastic and light up from inside. Chief Smoky shares its skirted body design with various other robots, including Ideal Toys' Robert the Robot and Matsudaya's Gang of Five.

▶ Bump-and-go

Although Chief Smoky did not walk on two legs, bump-and-go skirted robots like this had a sinister appeal as they roamed 1960s living rooms.

▲ Alternative name

This robot and its box are highly collectible items. On some models, the chest-plate reads "Mr. Chief."

Antenna is a delicate wire

Bulbs make the eyes glow as if on fire

Chest is decorated with valves and circuitry

Helmet emits smoke from a heated oil reservoir

Arms swing as the robot moves

Skirt hides the bump-and-go wheel mechanism

Specification: Chief Smoky

First manufactured:	1960s
Country of origin:	Japan
Manufacturer:	Yoshiya
Height:	12 in (30 cm)
Power source:	Battery-operated
Capabilities:	Changes direction, smokes, head lights up

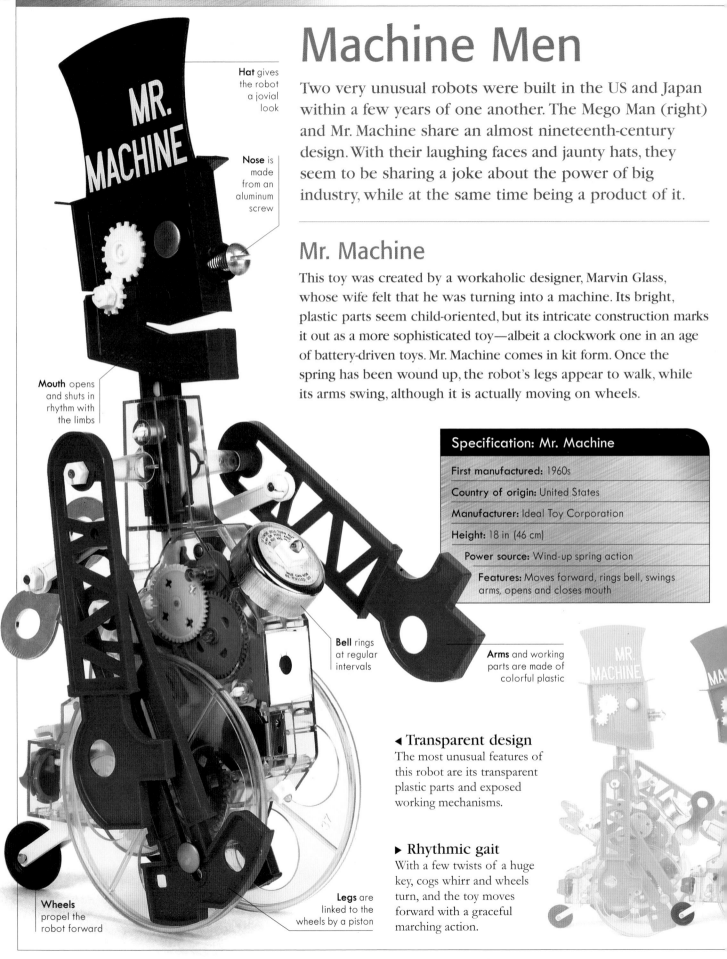

Hat gives the robot a jovial look

Nose is made from an aluminum screw

Mouth opens and shuts in rhythm with the limbs

Machine Men

Two very unusual robots were built in the US and Japan within a few years of one another. The Mego Man (right) and Mr. Machine share an almost nineteenth-century design. With their laughing faces and jaunty hats, they seem to be sharing a joke about the power of big industry, while at the same time being a product of it.

Mr. Machine

This toy was created by a workaholic designer, Marvin Glass, whose wife felt that he was turning into a machine. Its bright, plastic parts seem child-oriented, but its intricate construction marks it out as a more sophisticated toy—albeit a clockwork one in an age of battery-driven toys. Mr. Machine comes in kit form. Once the spring has been wound up, the robot's legs appear to walk, while its arms swing, although it is actually moving on wheels.

Specification: Mr. Machine
First manufactured: 1960s
Country of origin: United States
Manufacturer: Ideal Toy Corporation
Height: 18 in (46 cm)
Power source: Wind-up spring action
Features: Moves forward, rings bell, swings arms, opens and closes mouth

Bell rings at regular intervals

Arms and working parts are made of colorful plastic

◀ **Transparent design**
The most unusual features of this robot are its transparent plastic parts and exposed working mechanisms.

▶ **Rhythmic gait**
With a few twists of a huge key, cogs whirr and wheels turn, and the toy moves forward with a graceful marching action.

Wheels propel the robot forward

Legs are linked to the wheels by a piston

Mego Man

This rare and much-sought-after Japanese tin toy was the inspiration for Mr. Machine. Like Mr. Machine, it has a rhythmic, piston-driven walking action, and a laughing face. Although Mego Man comes preassembled, it includes a construction plan and operating instructions. The large wheels on either side are motor-driven, operating the arms and legs, and a red bulb in the toy's chest lights up as it walks. Much of Mego Man's design harks back to earlier Japanese toys, with its lithographed tinplate and rivets.

▶ Aesthetic robot

There are few, if any, more elegantly decorated toy robots. Mego Man's colorful geometric detailing on tinplate is reminiscent of early-twentieth-century artists such as Picasso and Matisse.

Specification: Mego Man

First manufactured: 1950s	
Country of origin: Japan	
Manufacturer: SY Toys	
Height: 18 in (46 cm)	
Power source: Wind-up spring action	
Features: Moves forward, swings arms, chest lights up, opens and closes mouth	

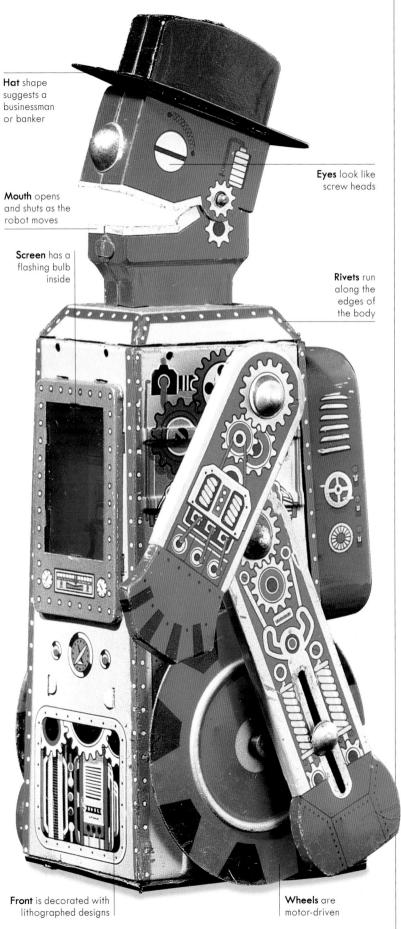

Hat shape suggests a businessman or banker

Eyes look like screw heads

Mouth opens and shuts as the robot moves

Screen has a flashing bulb inside

Rivets run along the edges of the body

Front is decorated with lithographed designs

Wheels are motor-driven

Head is transparent, revealing gears

Gears light up as they rotate

Arms can be moved manually

Alarm lights up on the head

Head lights up as the robot moves

MIGHTY ROBOT

Air vent imitates the working machinery

Skirted base conceals the bump-and-go action

Gear Robots

Gear robots first appeared in the 1950s and became the favored style of the Japanese manufacturers Horikawa and Taiyo. They were battery-operated, but a window exposed a whirring mechanism of colored gears, reminiscent of clockwork toys. When lit up and in motion, a gear robot is an impressive sight—a combination of industrial machine and Space Age toy.

Mighty Robot

A hybrid of classic robot styles, Mighty Robot is a real toy of the 1960s. Its illuminated transparent head makes it a striking and unusual robot. However, the body shape and lithographed detailing on the chest is identical to other toys such as Chief Smoky.

◀ Hidden mechanism
Mighty Robot's skirted base hides a pivoted, wheel-based design, which propels it and changes its direction when it hits an obstacle.

Specification: Mighty Robot	
First manufactured: Late 1960s	
Country of origin: Japan	
Manufacturer: Yoshiya	
Height: 11¾ in (30 cm)	
Power source: Battery-operated	
Features: Changes direction, gears turn, head lights up	

Blink-A-Gear

This large robot has an austere black body highlighted in bright red, and a backlit chest full of gears that spin as the robot walks. Its green eyes light up, making it one of the most powerful-looking of the robots. Blink-A-Gear is humanoid in shape, with legs instead of a skirt, echoing the look of earlier robots. However, the plastic elements anticipate later decades of toy design.

▶ Exacting work
The robot's body is elegantly finished, indicating a high level of manufacturing precision by Taiyo. Many of their toys are still in very good condition.

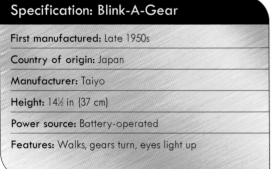

▲ Collector's item
Blink-A-Gear is a favorite of collectors because it combines the charm of earlier toys, such as Robby and Zoomer, with the inventiveness of gears.

Specification: Blink-A-Gear

First manufactured:	Late 1950s
Country of origin:	Japan
Manufacturer:	Taiyo
Height:	14½ in (37 cm)
Power source:	Battery-operated
Features:	Walks, gears turn, eyes light up

Eyes light up for striking effect

Headset feature is typical of other, smaller gear robots

Gears are seen turning from the front and sides

Hands are shaped like wrenches

Legs are inspired by Robby, making it a post-1955 toy

Feet conceal rollers to assist walking action

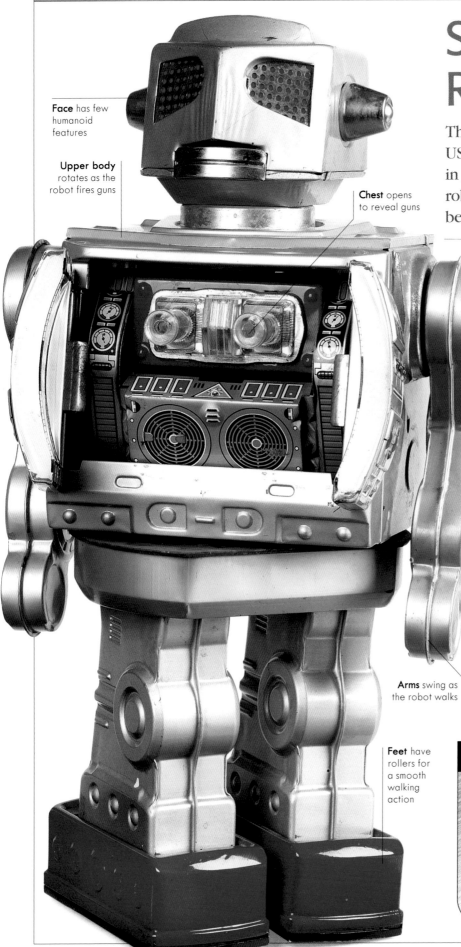

Face has few humanoid features

Upper body rotates as the robot fires guns

Chest opens to reveal guns

Arms swing as the robot walks

Feet have rollers for a smooth walking action

Shooting Robots

The space race between Russia and the US inspired manufacturers, particularly in Japan, to promote fearsome space robots, suggesting that the Cold War was being fought for control of outer space.

Super Space Giant

Powerful-looking and unusually large—the robot weighs 5 lb (2.25 kg)— Super Space Giant lives up to its name. The robot strides forward a few paces, then pauses. Doors in its chest spring open and ray guns emerge. The upper body rotates as the guns light up and emit firing noises, then the doors close, and the cycle repeats itself. This robot has large, screened eyes, but some variations feature an astronaut's face within the robot's head.

◄ **Imposing toy**

Super Space Giant is largely of tinplate construction, and is equipped with plastic guns. Apart from its large size, the basic design is similar to other battery-operated Horikawa space and piston robots. The rotating upper body is a unique feature.

Specification: Super Space Giant
First manufactured: 1960s
Country of origin: Japan
Manufacturer: Horikawa
Height: 16 in (41 cm)
Power source: Battery-operated
Features: Walks, rotates upper body, chest opens, guns light up and emit firing noises

Attacking Martian

After striding forward with arms swinging, Attacking Martian stops and a double-barreled gun appears from his chest, lighting up and emitting a machine-gun noise. Then the gun retracts and the robot strides on. This battery-powered toy is one of many Attacking Martians by Horikawa.

▶ Proud warrior

Attacking Martian's robustly constructed body is made of tinplate, while the eyes, guns, and arms are made of plastic. It is one of the most strikingly designed and attractive robot toys of this period, with intricate, colorful, lithographed rivet and dial details.

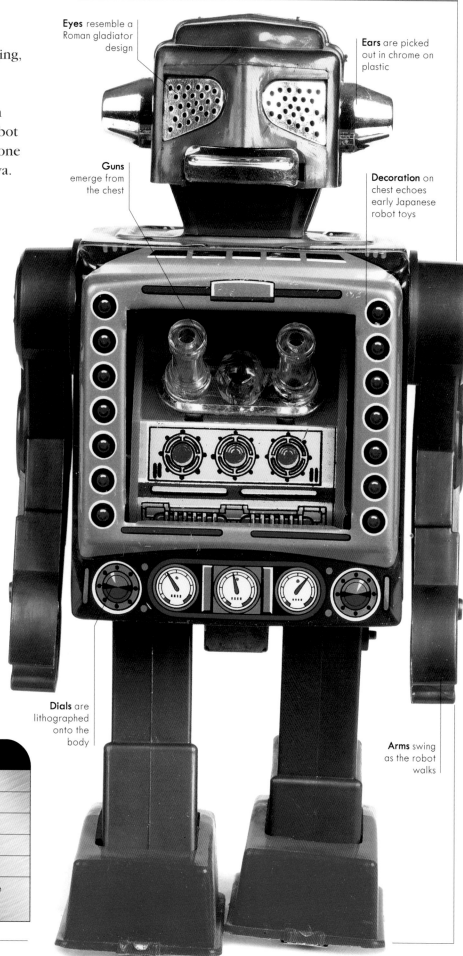

Eyes resemble a Roman gladiator design

Ears are picked out in chrome on plastic

Guns emerge from the chest

Decoration on chest echoes early Japanese robot toys

Dials are lithographed onto the body

Arms swing as the robot walks

▲ Fierce explorer

The evocative box art shows Attacking Martian to be gigantic and terrifying as it strides across a war-torn landscape.

Specification: Attacking Martian

First manufactured: 1970s

Country of origin: Japan

Manufacturer: Horikawa

Height: 9 in (22.5 cm)

Power source: Battery-operated

Features: Walks, guns light up and emit firing noise

Head is
shaped like a
Florentine
dome

Head unfolds
into three
segments

▲ **Flower power**
As Tulip Head walks, the
three metal petals in its
head open and the TV
screen and dome light
inside are illuminated. A
rotating camera adds to
the effect. The petals then
close and the action is
repeated once again.

Space Robot

Made in the 1960s, this unusual
robot was also known as "Tulip
Head," and it was one of the first
with an automatic transforming
action. Distinctive and colorful,
it is a distinguished toy in either
open or closed position. It is a
sought-after collector's piece
when in mint condition and
with its original packaging.

Early transformer

Space Robot is a tinplate toy with
considerable plastic detailing on its
eyes, dome, arms, collar, TV camera,
and screen. It walks forward with
sound blaring while its domed head
opens up like a flower, revealing
a rotating TV camera and screen.
On its chest is a depiction of nonfunctional controls, a tape recorder,
and its serial number, X-70.

▲ **Pop art**
Part of the joy of this robot
is its profusion of colors.
It could have come from
the Pop Art studios of
Andy Warhol. Even the
insides of the petals have
lithographed details.

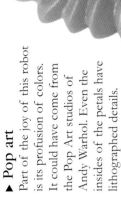

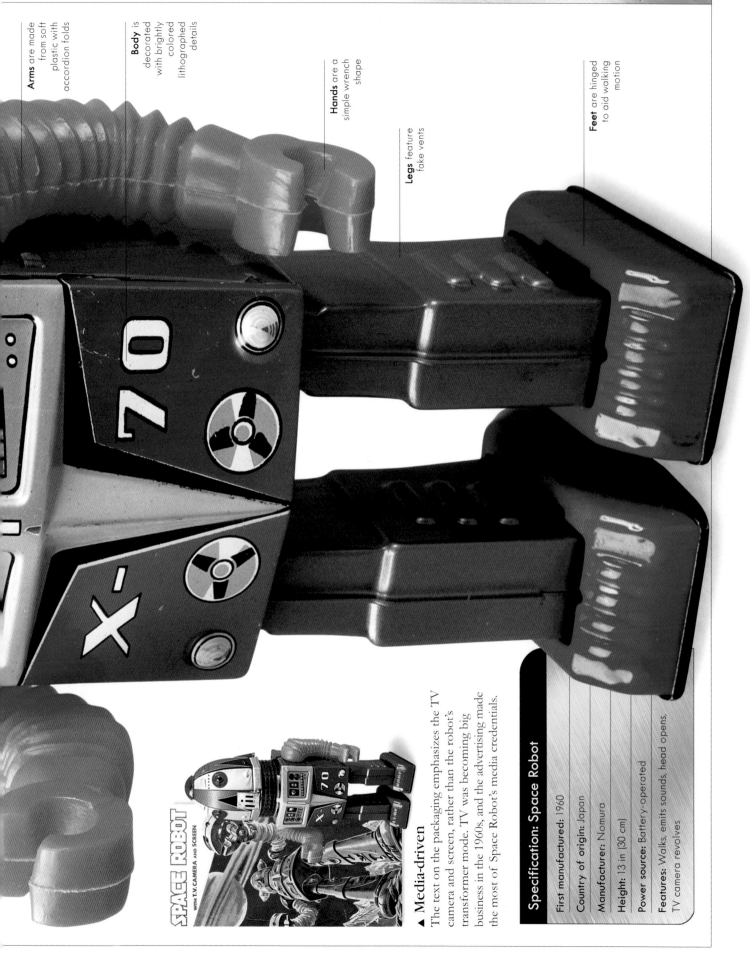

Arms are made from soft plastic with accordion folds

Body is decorated with brightly colored lithographed details

Hands are a simple wrench shape

Legs feature fake vents

Feet are hinged to aid walking motion

▲ Media-driven

The text on the packaging emphasizes the TV camera and screen, rather than the robot's transformer mode. TV was becoming big business in the 1960s, and the advertising made the most of Space Robot's media credentials.

SPACE **ROBOT**
WITH T.V. CAMERA AND SCREEN

Specification: Space Robot

First manufactured: 1960

Country of origin: Japan

Manufacturer: Nomura

Height: 13 in (30 cm)

Power source: Battery-operated

Features: Walks, emits sounds, head opens, TV camera revolves

Robby and friends

ROBBY THE ROBOT was not the first robot movie icon, but he was certainly the first robot to be as famous a toy as he was a film character. So great was his impact at his 1956 debut in *Forbidden Planet*, that he transcended his original role and went on to became a star in his own right. Toy makers clamored to jump on the Robby bandwagon—or space wagon.

▲ **Space Robbys**
These two robots, Moon Explorer from the 1950's (left) and Robby Robot from the 1960's (right), gave the popular tin toy astronaut a little Robby the Robot magic.

They say that imitation is the sincerest form of flattery, and Robby certainly had a legion of admirers. In his original film role as Dr. Morbius's assistant, his stove-like body, "Michelin Man"-style legs, whirring gears, flashing lights, and lovable personality made him the first robot celebrity. Indeed, Robby was the embodiment of Isaac Asimov's vision of the friendly, if enigmatic, machine.

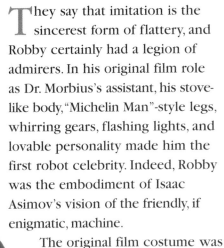

The original film costume was designed by Robert Kinoshita, who later designed the *Lost in Space* robot B9. Not only was B9 an obvious descendant of Robby, but it became another best-selling spinoff toy.

A marketing bonanza
Robby's purely superficial resemblance to a human being ensured that the public regarded the robot as a real, intelligent machine rather than as an actor in disguise. Robby soon became a

celebrity in his own right and he went on to feature in another film and to make guest appearances in several TV shows.

Before Robby, toymakers had been marketing tin men with no story to tell. Robby's appearance coincided with the rise of the first mass-marketable celebrities of

Head
shape recalls earlier toy designs

Legs are the most obvious reference to Robby

◀ **Planet Robot**
This 1958 Japanese robot toy by Yoshiya is a copy of Robby, except for the color and visor design.

▶ **Blink-A-Gear**
This, along with other 1960s gear robots, showed Robby's influence without being a direct copy.

popular culture. Owning a Robby
toy gave children their own mini
superstar, complete with the same
features as the real Robby.

There was a flood of Robby and
Robby-inspired toys when the film
was released, setting the standard
for later movie and TV spinoffs.

Japanese toys

Nomura of Japan made many of
the original toys, which are now
highly sought-after by collectors.
Few were known as "Robby," but
they were given names such as
"Mechanized Robot" (Nomura's
original Robby toy, first made in
1956), "Mechanical Robot," or
"Piston-Action Robot."

This was a challenge for other
toymakers who wanted to
capitalize on Robby's robotic
appeal but lacked official studio
approval. Soon, Robby imitators
were everywhere.

But not all Robby's relations
were cynical copies. The design
of robots such as Blink-A-Gear,
Thunder Robot, Space Robot, and
others simply acknowledged that
Robby had changed the direction
of robot design.

Leader of the pack

Once Robby had captured the
public's imagination, "robot toy"
conjured up a definite image of
whirring gears, a domed head,
flashing lights, a stocky body,
bulbous arms and legs, and,
importantly, a personality. For a
long time, "robot" meant "Robby."

Head retains
dome-like
shape

Arms are
forward-
pointing,
like a
sleepwalker's

◄ **Thunder Robot**
A popular toy that appeared in 1968,
Thunder Robot retained some basic
Robby features within a design unique
among robot toys. It is shown here in
a later reproduction.

Robby vehicles

The space wagon driven by
Robby in *Forbidden Planet* started
a craze for toy vehicles with a
Space Age twist, especially if
Robby or a Robby-style robot
was in the driver's seat.

Robby is
fixed to the
top of the
tank

▲ **Space wagon**
This Space Tank has a chassis design similar
to other push-and-go toy tanks of the time,
but the "Robby" turret cleverly recalls the
space wagon in the original movie.

▼ **Road roller**
Made in the 1950s, this rare lithographed
tin Road Construction Roller is driven
forward by Robby. Visible engine actions
include pumping and lighted pistons.

Propeller spins in a helicopter style

Head sits on top of the legs

Casing is designed to be ultra-strong

Panel flashes when the robot walks

Eyes light up and appear to spin

Mouth recalls car-style grille

Arms have concertina-style folds

Thunder Robot

This unusual tinplate toy from the 1960s marks a departure from the popular "tin man" design. This robot is effectively a "walking head" that launches an aggressive display of lights and noise as soon as it is switched on.

Action-packed toy

Thunder Robot marches forward in search of an opponent with rotor spinning, its eyes and head panel flashing danger. When it finds its target, it raises its arms to reveal armored hands that light up and emit a machine-gun noise for a few seconds. Then it lowers its arms and marches off again, in search of new adversaries.

Legs
are strong and
widely spaced
for stability

Body is made
of tinplate
and clipped
together

Feet are brightly colored in
contrast to the body

▶ Two-handed shooter

The original toy is now rare, especially in mint condition, but there are a number of reproductions available, which are mainly aimed at adult collectors. This version, made by Tim Tom, reveals Thunder Robot's flashing guns, which are built into its outstretched palms.

◀ On the attack

The original packaging shows Thunder Robot striding through a war-ravaged lunar landscape, laying waste to everything in its path using its trademark two-handed attack.

Specification: Thunder Robot

First manufactured: 1968

Country of origin: Japan

Manufacturer: Asakusa

Height: 11 in (28 cm)

Power source: Battery-operated

Features: Walks, raises arms and fires guns with sound accompaniment, lights flash, propeller rotates

Mr. Mercury

It is rare for a tinplate toy to be able to carry out simple tasks, but Mr. Mercury really was ahead of his time. He anticipated the industrial robots used for manual labor, and the training robots such as Armatron, which demonstrated the ability of a robot arm to grip and move objects.

Specification: Mr. Mercury	
First manufactured: 1960	
Country of origin: Japan	
Manufacturer: Marx, Line Mars Co. Inc.	
Height: 13½ in (34 cm)	
Power source: Battery-operated	
Features: Walks, bends at the waist, grips and lifts objects	

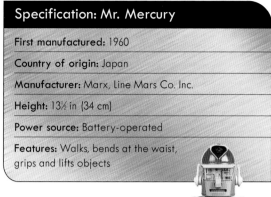

Space-mining robot

This handsome, remote-controlled robot can walk, raise his arms, bend at the waist to pick up an object, and move his arms inward and outward in a gripping action. His looks are impressive, too, with a red, winged helmet that recalls his namesake, and a nameplate mounted proudly on his chest.

▶ **Remote control**
The remote control unit is wired to the robot's body. Different buttons make it walk, stop, bend, and open or close its arms.

▼ **Versatile actions**
Mr. Mercury is able to grip and lift objects between his hands. First, he stands straight and opens his arms, then he bends while closing his arms until the object is within his grasp. The robot can then straighten up and walk while holding the object.

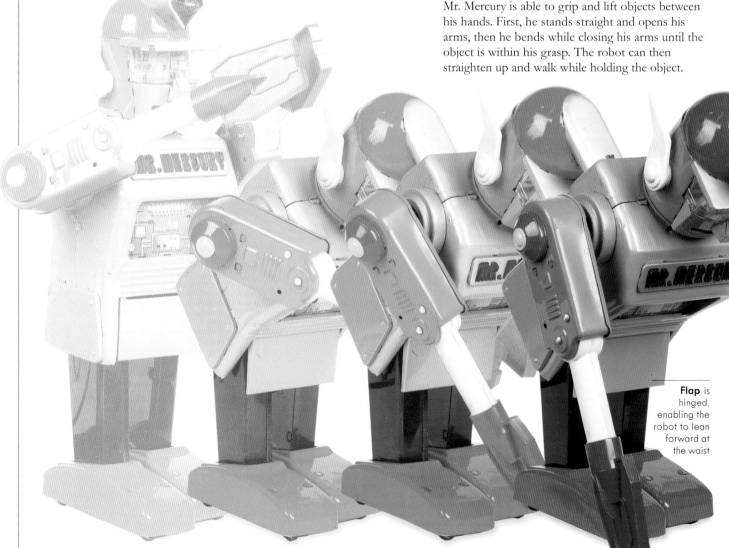

Flap is hinged, enabling the robot to lean forward at the waist

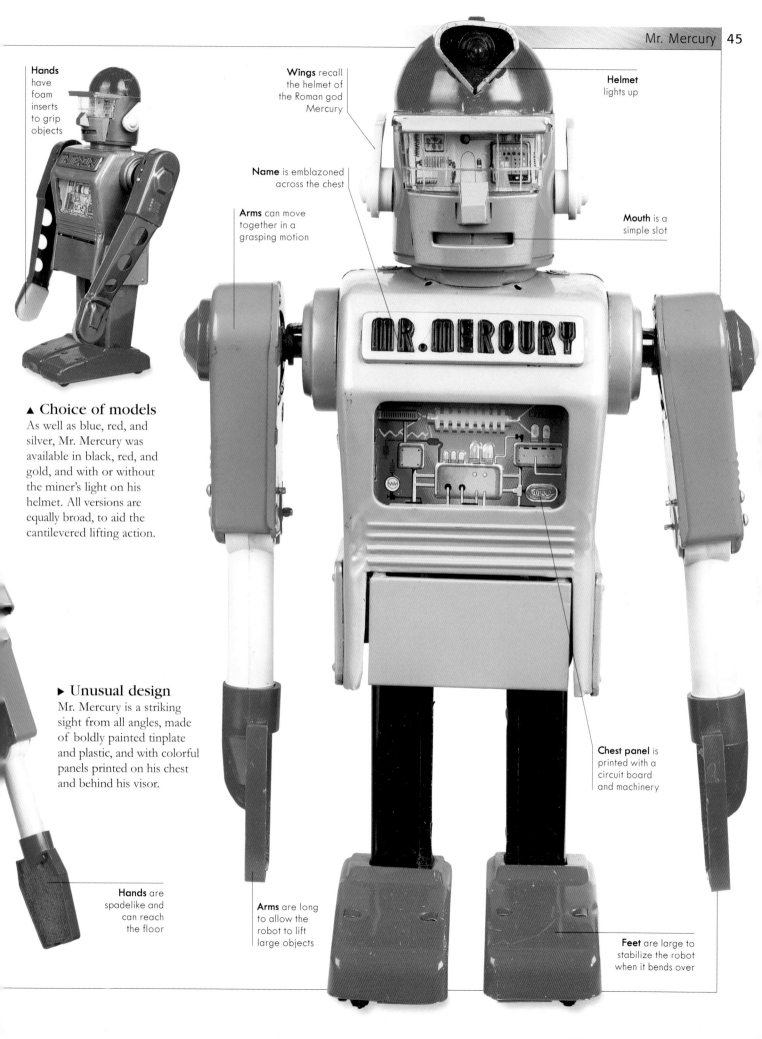

Hands
have foam inserts to grip objects

Wings recall the helmet of the Roman god Mercury

Helmet lights up

Name is emblazoned across the chest

Mouth is a simple slot

Arms can move together in a grasping motion

▲ **Choice of models**
As well as blue, red, and silver, Mr. Mercury was available in black, red, and gold, and with or without the miner's light on his helmet. All versions are equally broad, to aid the cantilevered lifting action.

▶ **Unusual design**
Mr. Mercury is a striking sight from all angles, made of boldly painted tinplate and plastic, and with colorful panels printed on his chest and behind his visor.

Chest panel is printed with a circuit board and machinery

Hands are spadelike and can reach the floor

Arms are long to allow the robot to lift large objects

Feet are large to stabilize the robot when it bends over

Engine Robot

In common with other piston- or engine-action robots, this toy has a basic configuration similar to earlier Smoking and Sparking Robots and later shooting Space Robots. The addition of pistons was probably inspired by Robby the Robot from the 1956 movie *Forbidden Planet*, and moved toy robot mechanics a step beyond toys with spinning gears.

Piston power

With the exception of its red metal pistons, this late-1970s robot is made of plastic. It has an interesting eye/nose configuration made from just one piece that resembles a silver car radiator grille. The detailing of Engine Robot, with its numerous rivets and flanges, was made possible by plastic casting. Engine Robot is an impressive sight as it moves forward on its rollers, pistons pumping and lights flashing. It also emits a realistic engine sound as it walks.

▶ **Imposing toy**
The artwork on Engine Robot's packaging, shown here in its original Japanese version, depicts the robot viewed from below, emphasizing its broad and bulky form as it towers over the rocky alien landscape.

Pistons
light up as the robot moves

Face is reminiscent of a radiator grille

Ears
resemble bolts

Shoulder is decorated with a large gear

Legs are flared, adding a stocky look

Wheels propel the robot forward

Elbows are fixed in bent position

Arms terminate in clawlike hands

► Imposing frame

The piston engine seems to burst out of the robot's chest. Broad shoulder gears add to its chunky appearance.

Specification: Engine Robot

First manufactured: Late 1970s

Country of origin: Japan

Manufacturer: Horikawa

Height: 8 ¾ in (22 cm)

Power source: Battery-operated

Features: Walks forward, pistons flash, emits engine sound

MachineRobo Series

With the MachineRobo series of 1982, Popy can be credited with the launch of robot toys that transformed into entirely different entities. Popy later became Bandai, and after two years, many of the MachineRobo figures—which also featured in a TV series—were launched in the US as the GoBot line. These were a different set of characters, but manufactured to the same specifications.

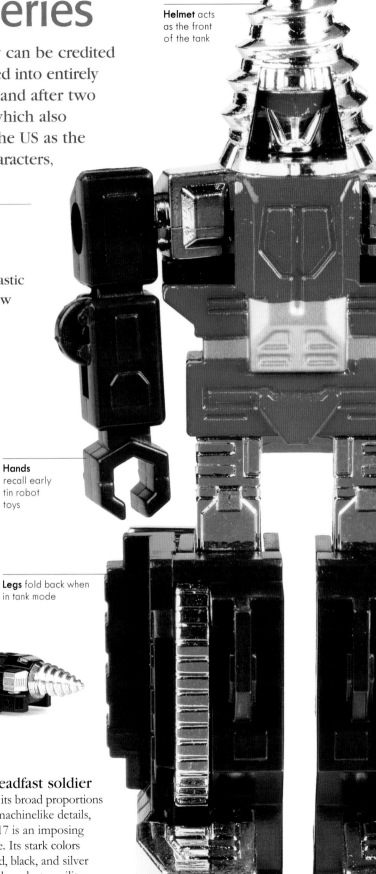

Helmet acts as the front of the tank

MR-17

Transforming from a robot warrior into a tank, MR-17 skillfully combines a die-cast metal structure and plastic detailing. For the US GoBot series, it was called Screw Head, and transformed into an "Enemy Robot Drill."

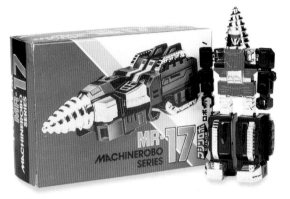

Hands recall early tin robot toys

▲ **Robot tank**
MR-17 is so cleverly designed that only the screw head and the treads appear to be shared by both the robot and the tank.

▶ **Streamlined tank**
The arms and legs of the robot have been folded and tucked into the tank's body, while the head becomes a boring or ramming weapon.

Legs fold back when in tank mode

Specification: MR-17

First manufactured:	1982
Country of origin:	Japan
Manufacturer:	Popy/Bandai
Height:	3½ in (9 cm)
Power source:	None
Features:	Transforms from robot to tank

▶ **Steadfast soldier**
With its broad proportions and machinelike details, MR-17 is an imposing figure. Its stark colors of red, black, and silver give the robot a military appearance.

MR-20

MR-20 transforms into a Porsche that can roll on its wheels; in robot mode, only the arms are movable. The robot's chest and helmet become the driver's cab and windscreen. As a GoBot, it was known as Crasher—a Renegade robot.

▶ **Car with character**
The white car was supposed to be an evil female robot, but a male version was also available in black.

MR-21

When MR-21 is transformed into car mode, all the robot features disappear behind the big windshield, spoiler, heavy wheels, and sloping hood. As a GoBot, it was known as the Renegade character Spoiler.

▶ **Sleek switchover**
The bulky figure of MR-21 robot transforms almost seamlessly into a sleek red Lamborghini Countach.

MR-40

A naval helicopter is changed into MR-40 as the back rotor splits to form the feet of the robot and the cockpit splits to become the arms, revealing the robot's face. In the US, MR-40 was known as Flip Top, a Guardian GoBot.

▶ **Flight of fancy**
MR-40 transforms into a blue Sikorsky Blackhawk. However, as a robot, it still appears fairly recognizable as a helicopter.

GoDaiKin Robots

The GoDaiKin series of transformable robots were a high point in the development of Japanese robot toys, both in their manufacturing precision and in their popular impact. They were die-cast, which enabled these high-quality, detailed robots to be made at a relatively low cost. Bandai's GoDaiKin were design classics in every sense.

Godsigma

Resembling a formidable shogun warrior, Godsigma was partly inspired by the TV and comic characters that proliferated in Japan from the 1960s to the 1980s. Most of the GoDaiKin robots transform into more than one articulated figure. The manufacturing techniques used would become the mainstay of the Transformers and GoBots that followed. Godsigma can be broken up to create three separate robots. The legs detach and transform to become two robots that are different in color and detail. The third warriorlike robot is a shortened version of the complete toy.

Specification: Godsigma

First manufactured: 1982

Country of origin: Japan

Manufacturer: Popy/Bandai

Height: 10½ in (27 cm)

Power source: None

Features: Transforms into three separate robots

Arms rotate at the shoulder

Head recalls a Japanese shogun warrior

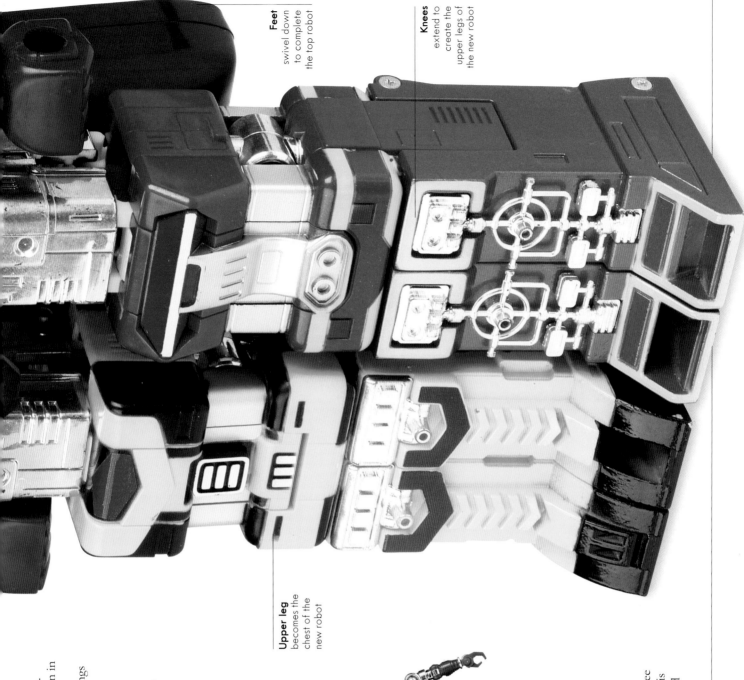

Feet
swivel down to complete the top robot

Knees
extend to create the upper legs of the new robot

Upper leg
becomes the chest of the new robot

▲ Simple transformation

Die-casting allows for complex construction. When Godsigma's legs are detached, a button in the back of each allows the upper leg to be extended. This in turn frees a panel that swings out to reveal articulated arms.

▲ Three-in-One

Godsigma can be broken down to create three separate robots: Thunder King (top), which is a shogun-style warrior, Earth King (left), and Ocean King (right).

Transformers

These transforming classics were not just collectible playthings, but represented a whole world of stories. Launched with the catch phrase "Robots in disguise," they spawned a TV series, which increased their popularity. From 1984 to 1986, Hasbro and Tonka Toys offered dozens of characters, and in 1986 the *Transformers* movie was released.

Optimus Prime in robot mode

◀ **From robot to truck**
Optimus Prime is a grand transforming conception. Changing from robot to truck to Autobot headquarters is not just a story-telling trick, but also a triumph of toy engineering, mixing die-cast metal with plastic to create a range of inexpensive, durable, precision toys.

Optimus Prime

A central character in the Transformers legend, Optimus Prime transforms into the Autobot headquarters to play a vital role in the war with the evil Decepticons. Part of the appeal of this toy is the additional parts, which add to the play possibilities. A new version of Optimus Prime was released in 2004 to celebrate its twentieth anniversary.

Legs fold and weapons are detached...

The transformation is complete...

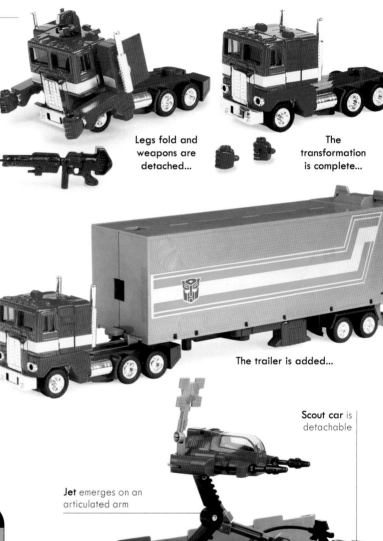

The trailer is added...

Scout car is detachable

Jet emerges on an articulated arm

Trailer opens to reveal the Autobot headquarters

▲ **Classic packaging**
The packaging shows the heroic Autobot Optimus Prime ready for action. The metallic-looking Transformers logo became almost as famous as the toys themselves.

Specification: Optimus Prime

First manufactured:	1984
Country of origin:	Japan
Manufacturer:	Hasbro/Takara
Height:	6½ in (16.5 cm)
Power Source:	None
Features:	Changes from a robot into a truck, which then has a trailer that houses robot headquarters attached to it

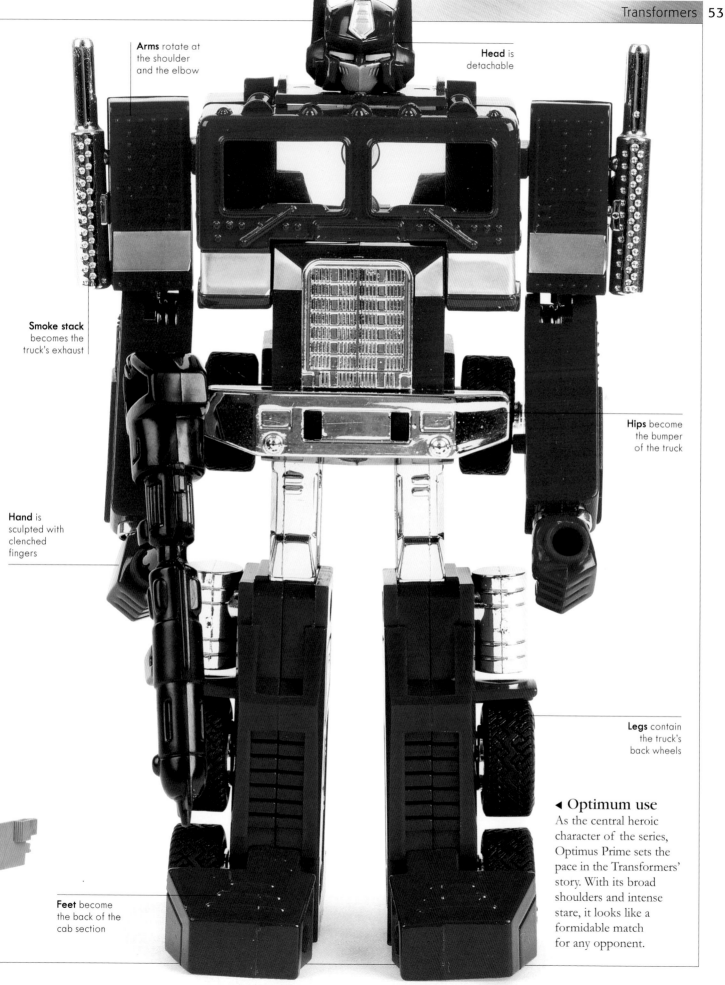

Arms rotate at the shoulder and the elbow

Head is detachable

Smoke stack becomes the truck's exhaust

Hand is sculpted with clenched fingers

Hips become the bumper of the truck

Legs contain the truck's back wheels

Feet become the back of the cab section

◄ **Optimum use**
As the central heroic character of the series, Optimus Prime sets the pace in the Transformers' story. With its broad shoulders and intense stare, it looks like a formidable match for any opponent.

Skywarp

In the Transformer world, Skywarp is an evil figure that battles against the good Autobots. The toy comes with twin missile launchers in both the jet and robot mode. Skywarp has interchangeable hands, weapon wings, tail assembly pieces, and detachable landing gear, which makes the robot quite distinct from the jet. Skywarp is one of a range of F-14 jets that make up the Decepticon Transformer series.

▶ **Front view**
Formidable wings and weapons make Skywarp a very imposing figure. The tiny head contrasts strongly with the bulky body of the robot. The shoulders are hinged, allowing the arms to be lifted up, and the hips swing, so Skywarp can be moved forward.

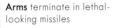

Arms terminate in lethal-looking missiles

Legs are jet thrusters

Spurs become the tail of the jet fighter

Weapons clip to the sides of the arms

Wings give the robot a sinister batlike appearance

▲ Back view
From behind, Skywarp is obviously a fighter jet in disguise. The shape of the wings is outlined in purple and white, the purple Decepticon logos are clearly visible, and the legs and feet are jet thrusters and a tail assembly.

Tail transforms into a leg extension

Specification: Skywarp

First manufactured:	1984
Country of origin:	Japan
Manufacturer:	Hasbro/Takara
Height:	8 in (20 cm)
Power Source:	None
Features:	Transforms from a robot into a jet fighter

Vents are printed and add detail to the design

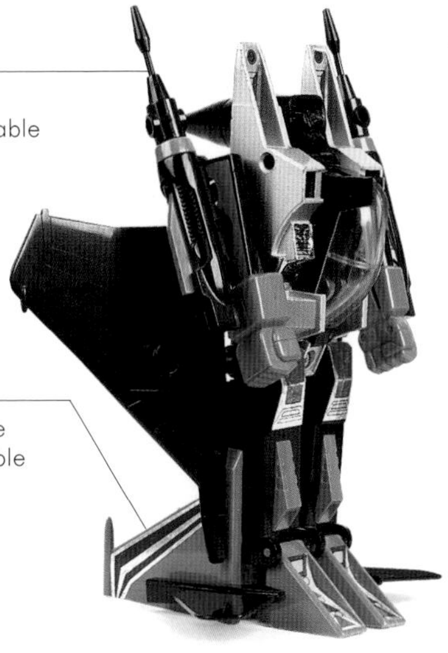

Missiles are detachable

Tail fins are clearly visible

Skywarp in robot mode

◄ From robot to jet

In transformation, Skywarp first loses its missile launchers and fistlike hands. Next, its feet flip up, the tail wing flips down, and the wings come off and reverse. Then the arms go in and the nose cone folds down through the body and straightens out. The landing gear, wings, and weapons are finally reattached to complete the change from robot to jet.

The weapons and fists are detached...

Fists are detachable

The nose cone is ready to swing forward...

Body is laid face-down

Legs form the rear of the aircraft

The transformation is complete

Wings are turned around and reattached

GoBots

Launched at the same time as Hasbro's Transformers, the GoBots were Tonka's contribution to the transforming craze. The Guardian and Renegade robots, and their many adventures on the planet GoBotron, were also made into a syndicated cartoon series.

Tank

This Renegade destroyer warrior is a clever combination of tank and robot, which both manage to look very different in their final forms. Only the weapons are recognizable in both robot and tank mode.

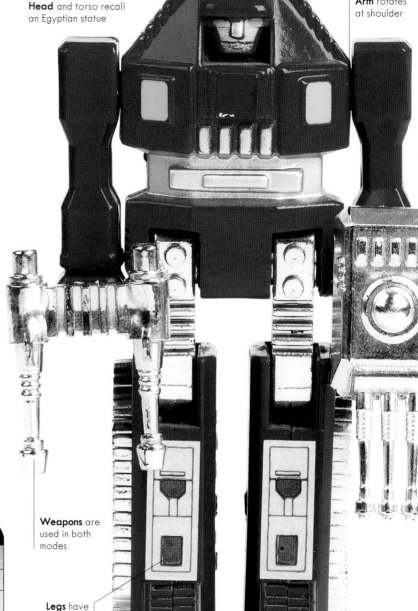

Head and torso recall an Egyptian statue

Arm rotates at shoulder

Weapons are used in both modes

Legs have stickers printed with pistons

▶ **Front view**
The robot's metal upper legs and weapons contrast with plastic lower arms, legs, and body. The back of the head becomes the tank's cabin and its legs the tractor treads.

Specification: Tank

First manufactured:	1984
Country of origin:	United States/Japan
Manufacturer:	Tonka Toys/Bandai
Height:	3¼ in (8.2 cm)
Power source:	None
Capabilities:	Transforms from tank into robot

▶ **From robot to tank**
This oblique view shows the tank treads, which are key to the transformation. The weapons are detached, then the robot is bent into a sitting position and flipped over. The weapons are then reattached to complete the switch to tank mode.

Tank in robot mode

Weapons are detachable

The tank folds over and loses its weapons...

Treads are visible on the robot's legs

The arms fold down and the legs lower...

Weapons are reattached

The transformation is complete

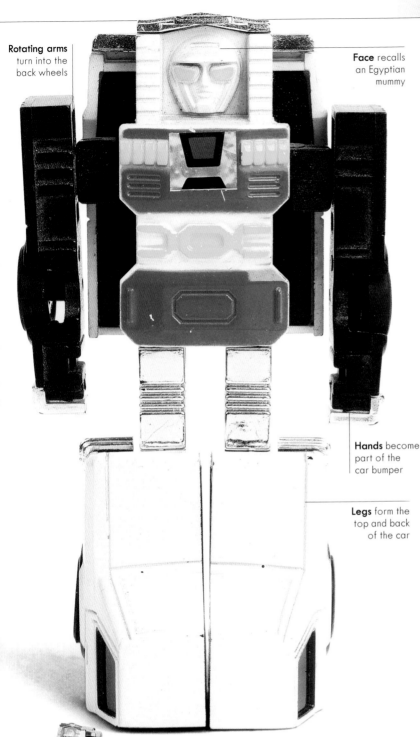

Rotating arms turn into the back wheels

Face recalls an Egyptian mummy

Hands become part of the car bumper

Legs form the top and back of the car

Hans-Cuff

As a friendly Guardian police car, Hans-Cuff can race to the scene of crime and become a police sergeant. As a robot, Hans-Cuff is still recognizable as a car, with giant feet that transform into the roof and trunk.

◀ Front view

The robot's feet form the back part of the car, while its arms become the forward wheel and bumper housing. Its low-slung arms are rotated from mid-torso, while the head is topped off by what in police-car mode is the bumper and license plate.

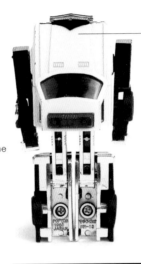

Back forms the car hood

◀ Rear view

Here we see clearly that the backpack is the windshield and hood of the car. As a car, however, Hans-Cuff gives no clue that it is capable of such a switch. The back reveals the die-cast metal leg assembly and the manufacturer's imprint, "Popy."

Specification: Hans-Cuff	
First manufactured: 1984	
Country of origin: United States/Japan	
Manufacturer: Tonka Toys/Bandai	
Height: 3½ in (8.7 cm)	
Power source: None	
Capabilities: Transforms from police car into robot	

◀ From robot to police car

Starting with the erect robot, the feet are turned under and the body tilted. The robot is then turned over and the arms folded to make the car body. The legs complete their turn and transform into the rear and trunk. Hans-Cuff is now ready for the road.

Hans-Cuff in robot mode

Head disappears under the car hood

The legs bend back and the body tilts...

Arms swivel to become the back of the car

The robot is turned over to reveal the front of the car...

Car now reveals no clue to its former shape

The transformation is complete

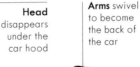

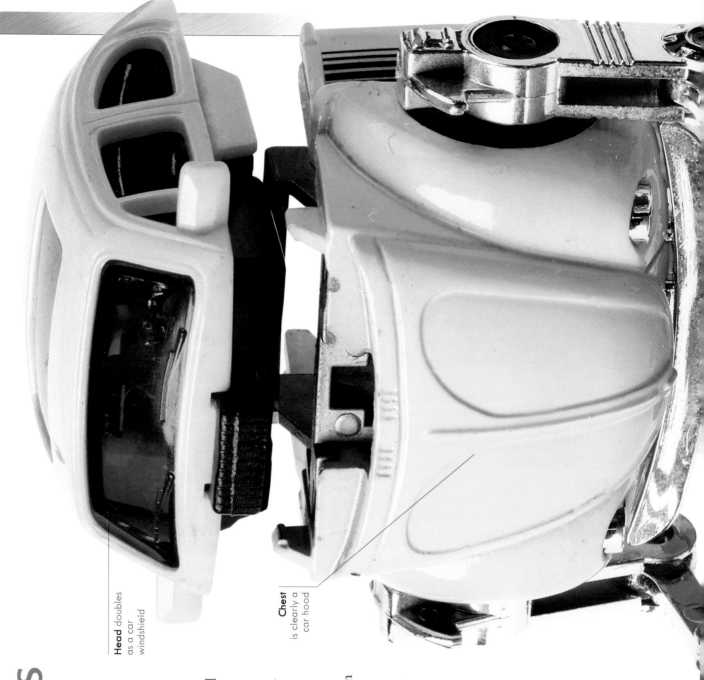

Head doubles as a car windshield

Chest is clearly a car hood

Super GoBots

Introduced by Tonka as the more powerful enemies of the GoBots, Super GoBots perhaps represent the marketing peak of this golden era of transforming toys. There are three series of Super GoBots: the first was issued with the original GoBots, the second series was larger, and the third is the rarest. In all, there are 19 Super GoBots, and all transform into cars.

Bug Bite

This is a Renegade robot from the first wave of GoBot toys, offering a curious combination of enemy robot and friendly- looking, yellow Volkswagen Beetle. In Super GoBot lore, Bug Bite, along with Herr Fiend, the archenemy of the GoBot Guardians, is known for its wickedness and cruelty. It stings its victims with deadly laser guns and also gives them the so-called Bug Bite "itch."

Specification: Bug Bite

First manufactured: 1984

Country of origin: United States

Manufacturer: Tonka Toys

Height: 5¼ in (14 cm)

Power source: None

Features: Transforms from robot to Beetle car

Legs made of die-cast aluminum fold out from beneath the car

Fender becomes part of the feet

Lower legs swing down to add height

Arms fold under the hood

Head becomes the car roof

Back wheels slot into main body

The front wheels emerge...

The head bends back and the hood moves forward...

The transformation is complete

The legs are tucked in...

Bug Bite in robot mode

▲ From robot to car

Bug Bite converts to a Volkswagen Beetle in smooth stages, but even in its robot mode, it clearly resembles a car. Despite its enemy credentials, it looks endearingly harmless in car mode.

▲ Durable design

The yellow body parts of Bug Bite are die-cast plastic, while the robot legs, internal structure, and wheel covers are of die-cast aluminum. This method of manufacture allows for high tolerance in the fitting of parts and greater accuracy in design, which could not be achieved by the traditional tin toys.

Cyber pets

EVER SINCE SPARKY the robot dog made his debut at the 1939 New York World's Fair, people have been fascinated by robot animals. Today, advances in robotic research have given cyber pets some of the intelligence and personality of real animals.

◀ Tekno New Born Pup

Nose lights up

This appealing puppy can walk, sit, and respond to voice commands. Its eyes light up, and it wags its tail. It comes with its own bone.

Real pets need a lot of looking after, but robotics has introduced a new generation of companions that need only a battery charge to keep them going. The chirping Furbies led the way in the mid-1990s, but it took Sony's appealing AIBO dog to convince people that cyber pets were more than just a turn-of-the-century gimmick. Today, there is a whole litter of robot dogs on the market, and there are plenty of robot cats for them to chase. While Sony's dog was a sophisticated

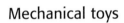

▲ MicroPets
These tiny toys respond to voice commands, and can walk and turn. Their eyes light up and their ears move.

robot aimed at adults, some of today's robot pets are really the latest in a long line of robot animal toys that stretches back as far as the 1950s. The main difference is that these state-of-the-art toys now have artificial intelligence on board.

Mechanical toys

The mechanics of robotic motion date back as far as the fourteenth century and include many animal automatons. These have evolved from simple clockwork mechanisms,

designed to entertain and amaze, into gyroscopically balanced machines. The arrival of microprocessors in 1971 helped to spur on the development of robot animal toys in the 1970s and 1980s, as did the miniaturization of sensors, motors, and other electronic components. The arrival of the personal computer then led to the development of robots that owners could program themselves.

Ears swing as the robot walks

Eyes light up with glowing LEDs

▲ Tomy Human Dog
A 16-bit CPU allows this dog to react to voices, walk, sit, tilt its head, and play tunes. It can develop one of 16 personalities to suit its owner.

◀ Robo Cat
This remote-controlled cat has a 750-word vocabulary and an array of sensors to respond to its environment.

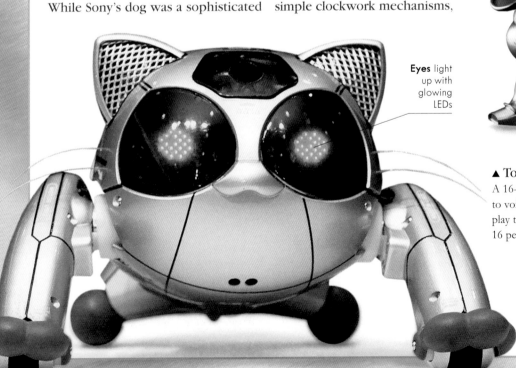

Today's advanced robots can walk and run independently, and many of these technologies are now being used in inexpensive cyber pets.

The main impetus behind the explosion of cyber pets and low-cost personal robots has been the influence of Japanese electronics companies such as Sony, as well as automotive companies such as Honda and Toyota, which are involved in developing a range of robots for use in the home. The size of these companies' investment in terms of both time and money suggests that personal robots, and therefore robot pets, have an interesting future.

▲ **Space-age dogs**
Space Dog, left, is a 1950s toy, while the modern Mega-Byte has motion detectors and can act as a guard dog.

back recorded sounds. In fact, the real key to the success of robot pets is their responsiveness. Most of them are now able to interact with their owners via sensors that detect movement, light, and sound. Many also have the ability to develop their own "personality" as they gradually learn which actions provoke a positive response from their owners.

Although cyber pets may not yet be as cuddly as their living counterparts, these robots are able to mimic the movements and even the sounds of the real animals. With their often endearing looks, cyber pets are becoming faithful companions to a willing public.

Companions of the future

Toys have been able to talk since Thomas Edison's speaking dolls appeared in 1886. Well over a century later, cyber pets can now actually communicate with their owners rather than simply play

Underwater robot pets

The craze for robot pets is not restricted to land-based animals, there is now a variety of aquatic creatures swimming onto the market. Their internal workings are similar to those of their walking relatives, but they have to be protected with a watertight casing.

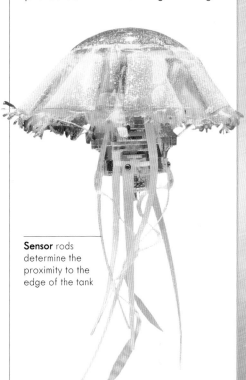

Sensor rods determine the proximity to the edge of the tank

▲ **Takara Aquaroid Jellyfish**
As this robot jellyfish moves up and down in the water, its internal lights glow. It can be set to swim at different speeds.

Propeller sits between the body and tail

Body houses batteries and a motor

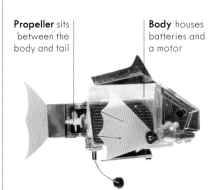

▲ **Takara Aquaroid Fish**
This aquarium pet swims by undulating its tail and fins—albeit with the aid of a small propeller. It is light-sensitive and can swim at two different speeds.

Ears move just like those of a real dog

Eyes light up to respond to owner

◄ **Tekno Dog and Cat**
The Puppy and Kitty in this matched set have distinct canine and feline personalities. One barks, and the other meows. Both walk, move their heads and tails, and express emotions through light and sound.

Pino DX

This large humanoid toy is named after Pinocchio, the puppet who wanted to be a real boy. It can be taught to walk, play games, dance, guard a room, and talk to other Pinos. As its owner trains the robot, it progresses through three phases of development. Pino is part of a trend in robots that learn through interaction with their owners and their environment.

Robots that grow up

Pino DX is a simplified, scaled-down version of a life-sized humanoid of the same name. Like its big brother, the toy can be programmed via a PC. It can teach itself through trial and error, and eventually becomes "friendly," "shy," or "naughty," depending on how its owner has treated it. A smaller, less sophisticated version of Pino DX, the Pino Interactive Robot Friend, is also available. Both versions of the toy can move their arms, show emotions, react to light and sound, and communicate via beeps and musical tones.

▼ **Modular frame**
Under the outer skin, Pino DX has a complex frame controlling the robot's movement. The skeleton is made up of 600 different components, which combine to make 26 modules. These allow Pino DX to walk, and even

▼ **Learning process**
Pino's publicity claims that the robot crash-landed on Earth and lost its memory, so it needs nurturing in order to learn and develop. Sensors in the head, hands, and face allow its owner to interact with Pino and respond to its moods and requests. Its "grown-up" personality depends on how its owner responds to it while it is learning.

Ear sensor activates games when pressed

Head houses a voice-recognition sensor

Visor changes color according to "mood"

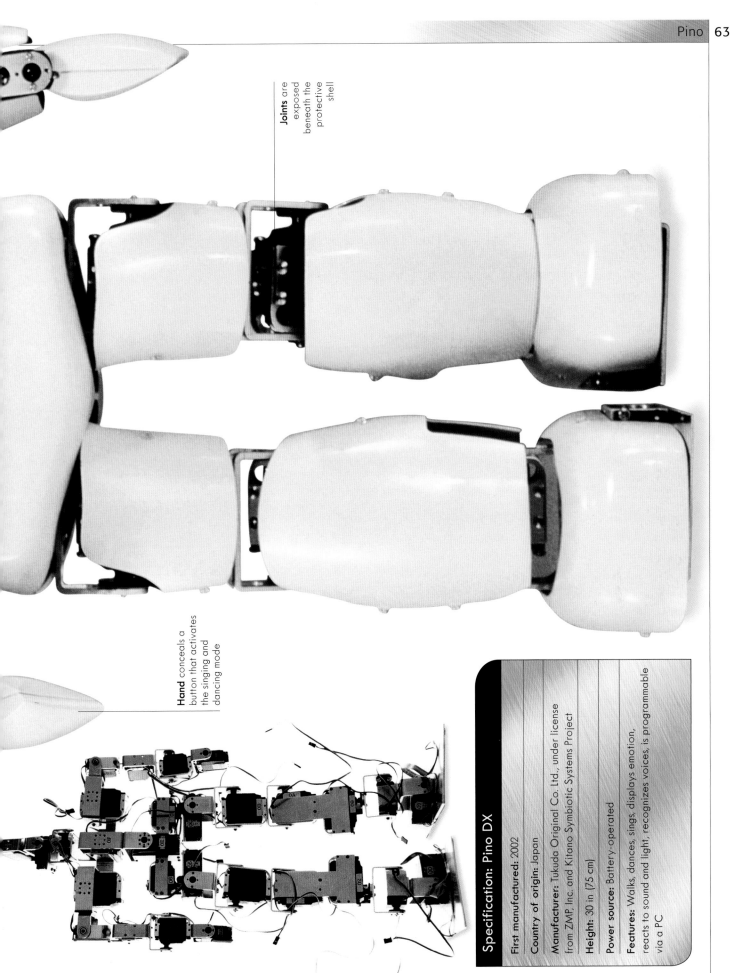

Joints are exposed beneath the protective shell

Hand conceals a button that activates the singing and dancing mode

Specification: Pino DX

First manufactured: 2002

Country of origin: Japan

Manufacturer: Tukuda Original Co. Ltd., under license from ZMP, Inc. and Kitano Symbiotic Systems Project

Height: 30 in (75 cm)

Power source: Battery-operated

Features: Walks, dances, sings, displays emotion, reacts to sound and light, recognizes voices, is programmable via a PC

RoboSapien

This very entertaining humanoid robot toy is a product of a team that includes Mark Tilden, the inventor of BEAM (Biology, Electronics, Aesthetics, Mechanics) robotics, and Wow Wee Toys. BEAM robots are based on simple control circuitry and sensors, but RoboSapien is nevertheless a sophisticated robot. A cross between a trooper and a gorilla in appearance, RoboSapien also disco-dances.

Action man

RoboSapien is remote-controlled but programmable. It can run, dance, kick, talk, pick up objects, and throw things, including tantrums. RoboSapien has three demo modes: "disco dance," "Kung-Fu," and "rude behavior," along with "international caveman speech capability." It has no onboard computer, but can run for eight hours on its batteries—no mean feat for any robot. Reacting to both touch and sound signals, its Interactive Reflex System uses a total of six sensors strategically located throughout its body to detect obstacles.

▶ **Universal appeal**
With its rugged appearance and feisty personality, RoboSapien appeals equally to adults and children. Its fluid biomechanical movements make its actions appear more human than robotic.

Shoulders are extra-wide, like a football player's

Visor resembles an astronaut's helmet

Elbow has a flexible joint for easy movement

Right hand has three articulated fingers

Feet are large to help balance

Specification: RoboSapien	
First manufactured: 2004	
Country of origin: Hong Kong	
Manufacturer: Wow Wee Toys Ltd.	
Height: 14 in (34 cm)	
Power source: Battery-operated	
Features: Runs, dances, swears, picks up and throws objects, and performs martial arts	

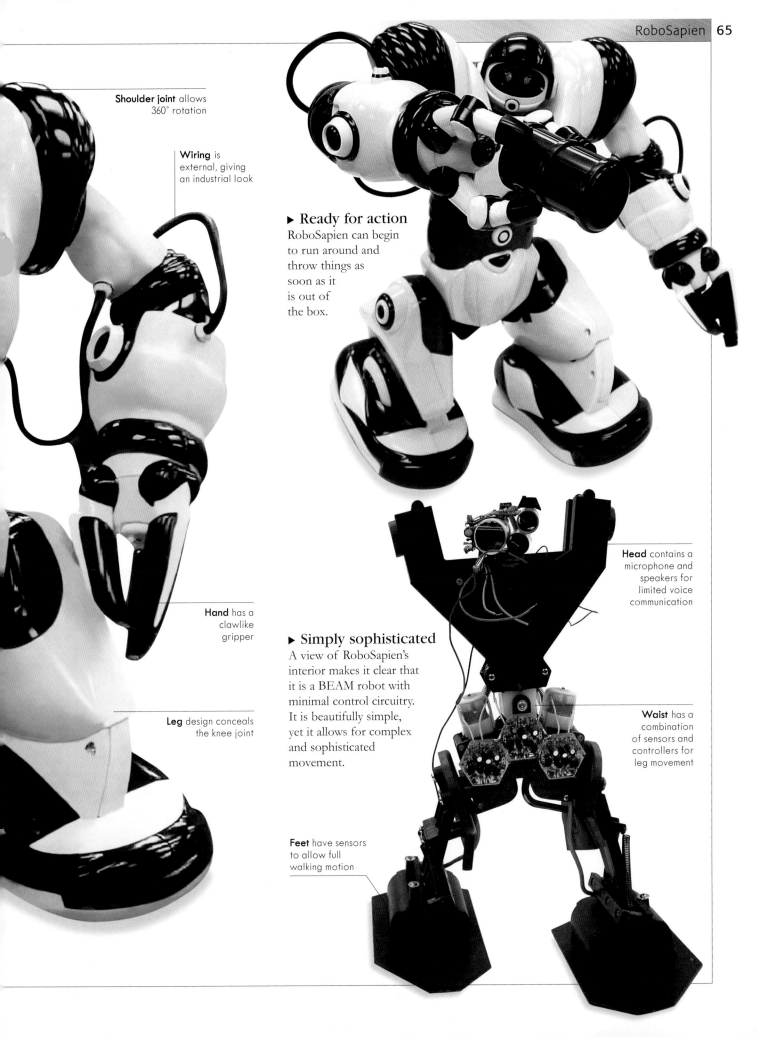

Shoulder joint allows 360° rotation

Wiring is external, giving an industrial look

Hand has a clawlike gripper

Leg design conceals the knee joint

▶ **Ready for action**
RoboSapien can begin to run around and throw things as soon as it is out of the box.

▶ **Simply sophisticated**
A view of RoboSapien's interior makes it clear that it is a BEAM robot with minimal control circuitry. It is beautifully simple, yet it allows for complex and sophisticated movement.

Head contains a microphone and speakers for limited voice communication

Waist has a combination of sensors and controllers for leg movement

Feet have sensors to allow full walking motion

AIBO Dogs

Sony's robot dogs walk, play, and can even obey voice commands. They build a relationship with their owners by learning to recognize them and communicate through lights, sounds, and gestures. The name of these intelligent toys means "companion" in Japanese.

AIBO ERS-7

This version of Sony's robotic dog responds to the sound of its name and recognizes its owner's face; its voice recognition technology allows it to react to 180 commands. ERS-7 walks more fluidly than earlier AIBOs. It matures from youth to adulthood, depending on how much the owner trains it. Paw, tail, and head sensors act along with specially designed software to direct its multiple motors.

▶ **Perfect pet**
This talented dog comes equipped with a PC card slot, memory stick media slot, digital camera, expressive LEDs, retractable headlights, stereo microphone, speaker, and touch sensor.

▲ **Playing ball**
Like the ERS-7, the earlier ERS-210 has an infrared sensor. When playing soccer, AIBO responds to touch and corrects its position using gyro sensors.

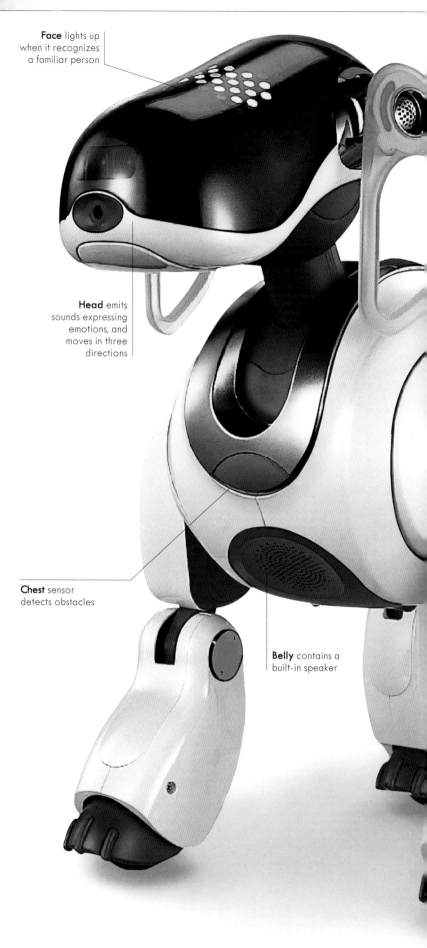

Face lights up when it recognizes a familiar person

Head emits sounds expressing emotions, and moves in three directions

Chest sensor detects obstacles

Belly contains a built-in speaker

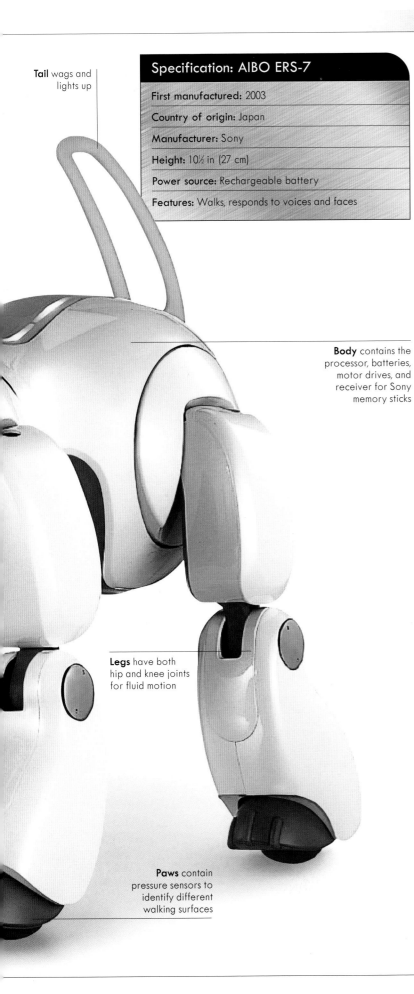

Tail wags and lights up

Specification: AIBO ERS-7

First manufactured: 2003

Country of origin: Japan

Manufacturer: Sony

Height: 10½ in (27 cm)

Power source: Rechargeable battery

Features: Walks, responds to voices and faces

Body contains the processor, batteries, motor drives, and receiver for Sony memory sticks

Legs have both hip and knee joints for fluid motion

Paws contain pressure sensors to identify different walking surfaces

The Sony kennel

The many types, sizes, and colors of AIBO pet dogs go back to earlier Sony research in mobility, communication, and control. Most are slight variations on the same model, while others, like the AIBO LM puppies, are unique but use the same technology.

Mouth incorporates a color camera

Nose contains an infrared distance sensor

▲ AIBO LM puppies

Puppies Latte and Macaron first appeared in 2001. They offer voice recognition, the AIBO tonal language, plus head, tail, and paw sensors. Their memory sticks can record and store photographs, and the puppies can communicate with other AIBOs.

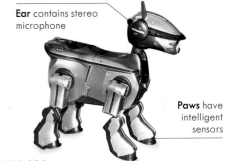

Ear contains stereo microphone

Paws have intelligent sensors

▲ ERS-220

This model, dating from 2001, has features such as a color camera, LEDs that display emotions, sensors in its head, tail, back, and paws, a PC card and memory stick, and a speaker in its mouth.

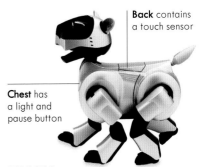

Back contains a touch sensor

Chest has a light and pause button

▲ ERS-210

The second generation AIBO was launched in 2000. It is a learning-mode robot that interacts with its owner. It has touch sensors, stereo microphones in its head, a speaker, a color camera, a memory stick, and a CPU.

GALLERY: Robot Toys & Collectibles

We have come to think of robot toys as mechanical friends, appealing to both children and adults. Although robot toys started as tin creations with simple mechanisms, today's machines reflect the complex world of twenty-first-century robotics. They come in all shapes, sizes, materials, and colors: some are in animal guise, some have their own vehicles, while others walk, roll, or even hop.

Back view

Plastic arms swing as the robot walks

Transparent body shows the motor mechanism

Tank treads are printed on to the base

Ranger Robot

Ranger moves its arms as it walks and makes engine noises. It is battery-operated and its mouth emits smoke.
- Date: 1955 ■ Country of origin: Japan
- Height: 10½ in (27 cm) ■ Manufacturer: Daiya

Robotank Z

The tinplate Robotank Z fires its guns and moves its arms as it rolls along. It is battery-operated, with a bump-and-go action.
- Date: 1960 ■ Country of origin: Japan
- Height: 10 in (26 cm) ■ Manufacturer: Nomura

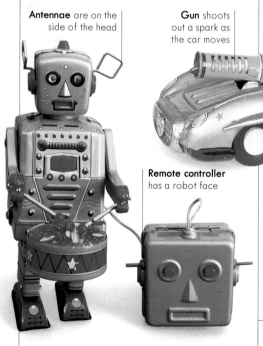

Antennae are on the side of the head

Gun shoots out a spark as the car moves

Robot is similar to ST-1 and Atomic Man

Remote controller has a robot face

Space Patrol

This is a friction toy with an internal flywheel. The robot is fixed inside the car; the wheels provide the driving motion.
- Date: 1955 ■ Country of origin: Japan
- Height: 5½ in (14 cm) ■ Manufacturer: Asahi

Body is tinplate

Robot Boat No.7

This floating boat is wound up by a crank handle. The robot seated inside is very similar to Zoomer.
- Date: Late 1950s ■ Country of origin: Japan ■ Height: 3½ in (9 cm)
- Manufacturer: ET Company

Windshield is made of plastic

Stern has a working propeller

Musical Drumming Robot

This tinplate Sparky lookalike walks while it beats its drum. It is battery-operated and works via remote control.
- Date: 1955 ■ Country of origin: Japan
- Height: 8½ in (21 cm) ■ Manufacturer: Nomura

Atlas Space Robot

This wind-up robot has its very own rocket. It is spring-driven and can walk as well as hop.
- Date: 1960s ■ Country of origin: Argentina ■ Height: 7 in (18 cm)
- Manufacturer: V2

Rocket can move independently on wheels

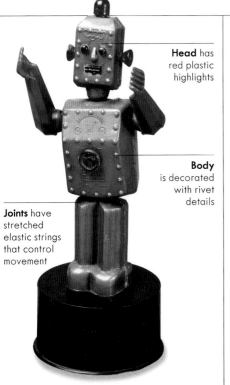

Head has red plastic highlights

Body is decorated with rivet details

Joints have stretched elastic strings that control movement

Head is of an unusual plastic box design

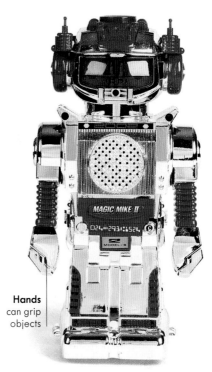

Hands can grip objects

Push-Up Robots

These molded plastic robots come in four different colors and can be made to move by pushing a plate in the base.

- **Date:** 1960 ■ **Country of origin:** Hong Kong
- **Height:** 4 in (10 cm) ■ **Manufacturer:** Unknown

Mechanic Robot

This battery-driven robot has a gear mounted in its chest, which revolves as it walks, turning its head from side to side.

- **Date:** 1970s ■ **Country of origin:** Japan
- **Height:** 12 in (30 cm) ■ **Manufacturer:** Horikawa

Magic Mike

This talking robot has a bump-and-go-action, his eyes flash as he moves around, and his mouth emits smoke.

- **Date:** 1984 ■ **Country of origin:** Japan
- **Height:** 12 in (30 cm) ■ **Manufacturer:** New Bright

Antenna is nonfunctional

Rivets are painted on the lithographed body

Chest contains a battery of guns

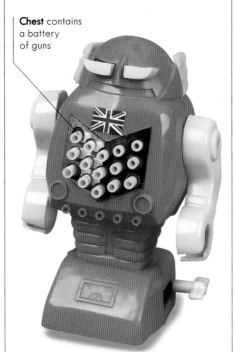

Visor recalls a *Star Wars* storm trooper

Body is made of molded plastic

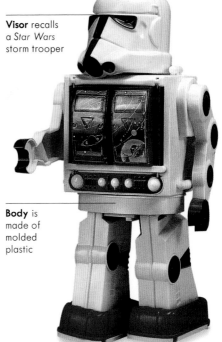

Hopping Robot

This jolly-looking clockwork toy has a bright, clownlike face. When wound up, it hops forward.

- **Date:** 1960 ■ **Country of origin:** Japan
- **Height:** 3¾ in (8 cm) ■ **Manufacturer:** Yoneya

Wind-Up Robot

This all-plastic robot rolls forward while firing the guns on its chest. It is decorated with the British flag.

- **Date:** 1970s ■ **Country of origin:** Hong Kong
- **Height:** 5½ in (14 cm) ■ **Manufacturer:** Unknown

Galaxy Warrior

This is a combination of a space robot and a storm trooper. As it walks, its chest opens, lights blinking, and it fires its guns.

- **Date:** 1979 ■ **Country of origin:** Japan
- **Height:** 12 in (30 cm) ■ **Manufacturer:** Amico

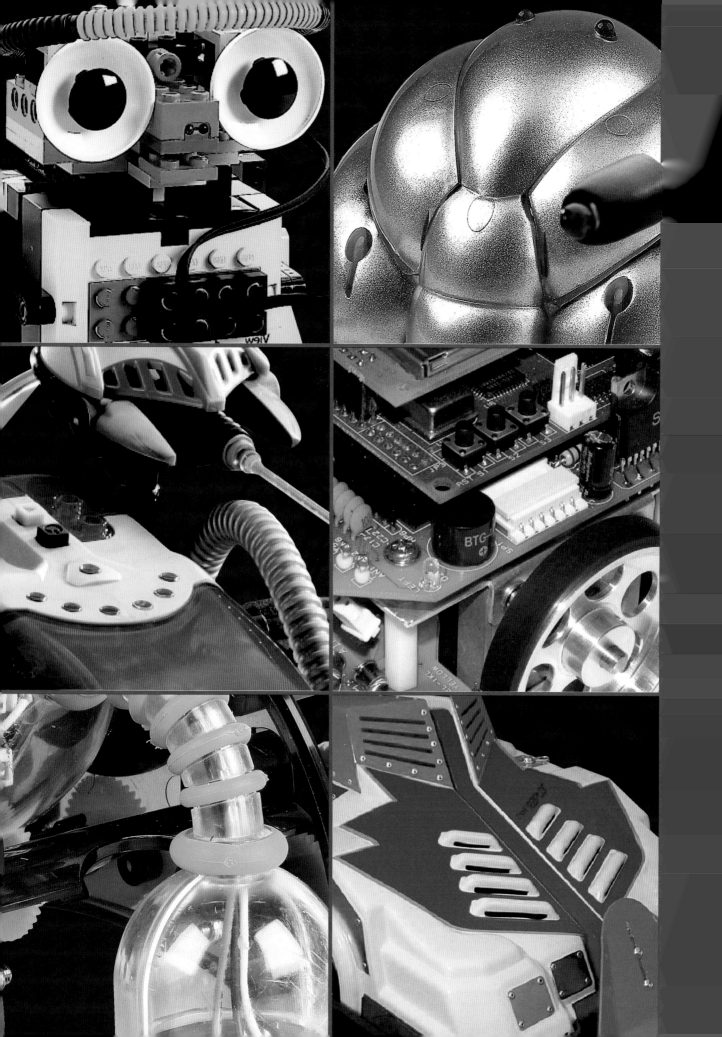

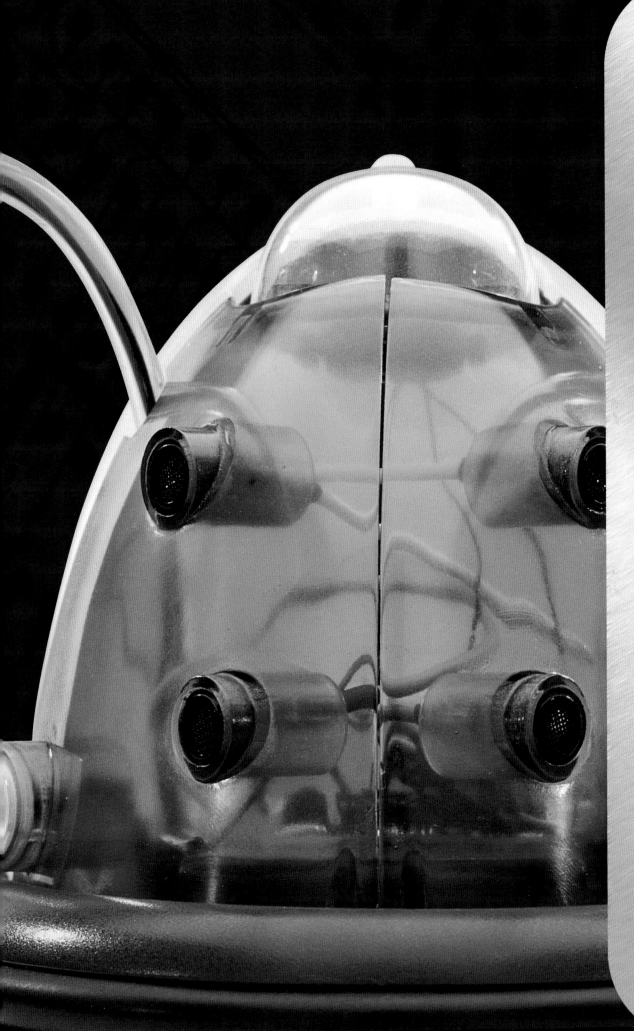

DIY Robotics

Hydrazoid
by iBOTZ

KITS BRIDGE THE gap between toys and the cutting edge of cybernetic research. Some kit robots allow hobbyists to pursue their interest for little cost; others are less affordable but offer students the key to the door of advanced humanoid robotics. While all robot kits have an educational element, they are also designed to be fun to build and operate.

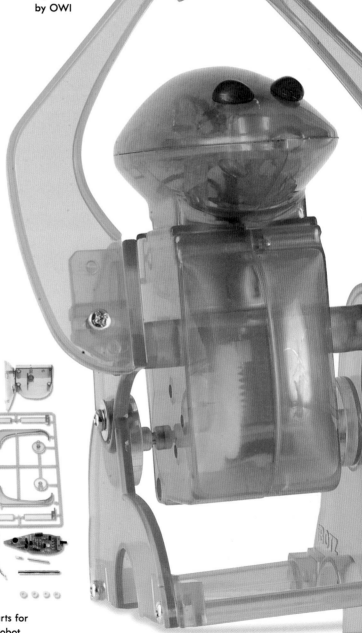

Jungle robot
by OWI

Whether they are designed for adults or children, all robot kits supply the building blocks and component technologies to make a robot. They demonstrate the basics of how robots work, how they are put together, and how they can be programmed. Once built, some robots carry out preset tasks, while others are able to learn via input from their owners.

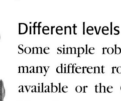

2-XL educational
toy by Mego

Different levels

Some simple robot kits, such as the many different robotic arms that are available or the CAM walking robot kits, demonstrate the complexities of movement and balance. More advanced kits, such as Dr. Robot's HR-6, stand at the higher end of the evolutionary scale, being sophisticated walking, talking, chess-playing, and Web-surfing companions.

But, above all, robot kits present the builder with the pleasure of "doing it yourself." At their simplest, kits provide the components to make rudimentary simulations of living beings: robot monkeys, mice, bugs, and even basic humanoids that use simple sensor arrays to react to sound and touch, follow light sources, or track the contours of a room. When they first appeared, simple kit robots such as these were a marketing response to the increasing public interest in robots, and also

The kit parts for
Jungle Robot

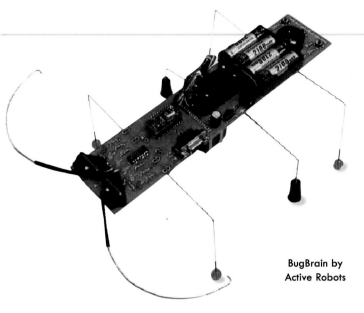

BugBrain by
Active Robots

Tribotz
by iBOTZ

the growing community of serious amateurs who were building their own robots specifically for international challenges and competitions.

Simple circuitry

BEAM (Biology, Electronics, Aesthetics, and Mechanics) technology, pioneered by Mark Tilden in 1989, enables kit-builders to create robots with simple onboard electronics that perform tasks such as moving toward a light or avoiding obstacles. BEAM kits demonstrate that robotics is as much about precision engineering as it is about artificial intelligence. They operate without the use of microprocessors and do not need programming, and many are solar-powered.

Intelligent kits

Some more complex kits, such as Cybot, exhibit the beginnings of robotic intelligence. The robot can be upgraded with additional components to enhance its intelligence and functionality.

Toy manufacturer LEGO has moved into the world of robotics, offering a range of robot kits, aimed at children, that can be built one intelligent brick at a time. Their Spybotics kits allow the builder to become part of an online community, downloading stories, adventures, and secret missions, which are fun to take part in but also teach the builder about the rudiments of networked computing.

Rascal construction
kit by Robix

Early standard-bearers

However, robot kits are not new. In the 1970s and early 1980s, when there was an enormous surge of interest in all things robotic, kits such as RB5X, Topo, and Androbot offered the first opportunity to really own a robot. They could be considered the standard-bearers for the army of robot kits that have marched onto the market since. As technology progresses at a rapid pace, and people become more familiar with the science of robotics, who knows what robots we will build for ourselves in the future?

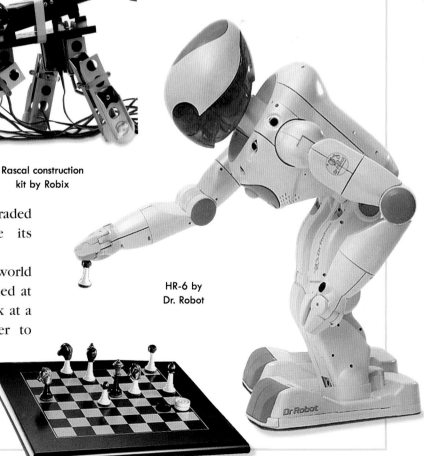

HR-6 by
Dr. Robot

Early Kit Robots

The popularity of the *Star Wars* droid R2-D2 and the emergence of personal computers and industrial robots in the early 1980s inspired the rise of personal robot kits: build-it-yourself, programmable robots that could move, communicate, and carry out simple tasks.

Topo

Topo is a play-oriented personal robot with a friendly, humanoid face. This remote-controlled kit connected to a personal computer was designed by Nolan Bushnell, founder of computer and games company Atari. Its predecessors were Androman, based on the Atari platform, and FRED (Friendly Robotic Entertainment Device). Later versions Topo II and III could talk. The last addition to the family had an onboard computer and was called BOB—Brains On Board.

▶ Computer-run
The original version of Topo runs fast on three wheels. Beneath its plastic exterior, a steel structure houses the drive, motor, and limited electronics. An Apple II or Apple IIe computer activates it remotely.

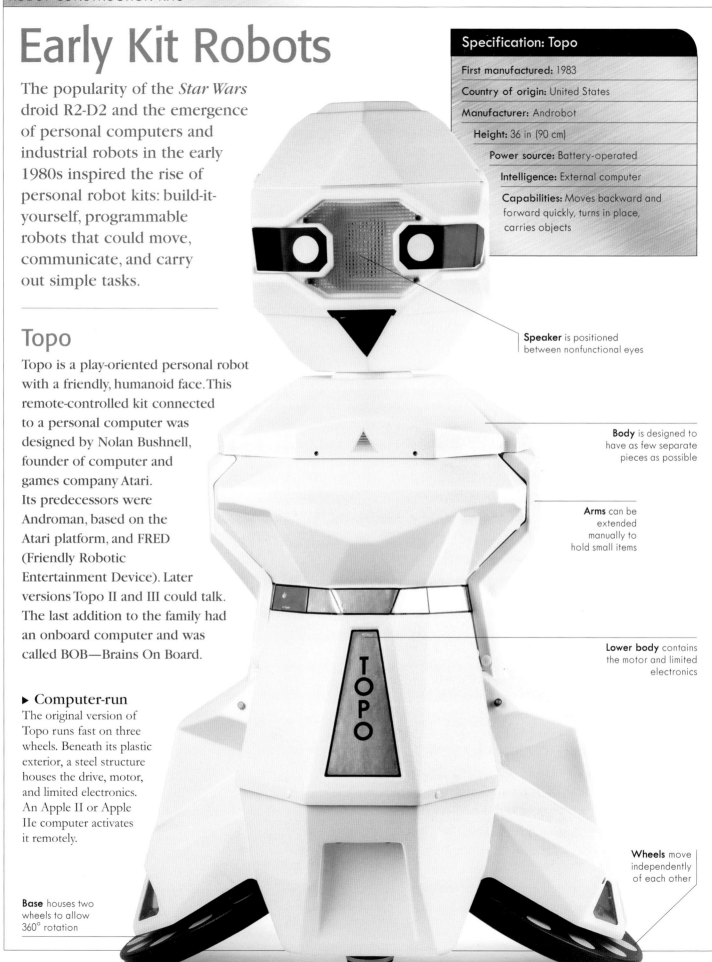

Specification: Topo
First manufactured: 1983
Country of origin: United States
Manufacturer: Androbot
Height: 36 in (90 cm)
Power source: Battery-operated
Intelligence: External computer
Capabilities: Moves backward and forward quickly, turns in place, carries objects

Speaker is positioned between nonfunctional eyes

Body is designed to have as few separate pieces as possible

Arms can be extended manually to hold small items

Lower body contains the motor and limited electronics

Wheels move independently of each other

Base houses two wheels to allow 360° rotation

RB5X

The first mass-produced personal robot for both educational and home use, RB5X has eight tactile sensors around its lower body and a bumper that allows it to sense when it comes into contact with another object. RB5X also recognizes when its batteries are low and will find its own battery recharger.

▶ Ahead of its time
RB5X has a microprocessor that was fast for its time, and a programmable Polaroid sonar sensor to detect objects in its path.

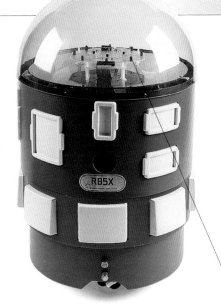

Sonar sensor is housed in the head

Specification: RB5X

First manufactured: 1982

Country of origin: United States

Manufacturer: General Robotics Corp.

Height: 23 in (59 cm)

Power source: Battery-operated

Intelligence: National Semiconductor 8-bit microprocessor

Capabilities: Moves, self-navigates, detects objects, recharges itself

Maxx Steele

A grown-up version of the Mr. Mercury toy, Maxx Steele can raise and lower its arms, and has rotating wrists and claws that can rotate and grip objects. Maxx comes with a wireless controller and can be programmed to talk, sing, play games, and give the time and date.

▶ Long arms
The robot's arms reach the ground, allowing it to pick up objects from the floor. Its chest contains an expansion slot for a cartridge that demonstrates its capabilities.

Shoulder rotation is controlled by a remote master controller

Specification: Maxx Steele

First manufactured: 1980s

Country of origin: United States

Manufacturer: Ideal Toy Company

Height: 23½ in (60 cm)

Power source: Battery-operated

Intelligence: Microprocessor 65C02 (CMOS)

Capabilities: Moves, talks, sings, picks up objects, plays games

Hero Jr.

The top part of this "guard" robot turns to detect movement, and can sound an alarm. It has a limited speech mode and can announce the time and date. Later models also have a working arm. It can perform self-diagnostics, and is equipped with a remote control.

▶ Friendly guard
Despite its guard function, Hero Jr. is not threatening. It is clearly a robot in the nonhumanoid style, and its plastic housing gives it the appearance of a friendly appliance.

Upper body contains programmable computing cards

Specification: Hero Jr.

First manufactured: 1984

Country of origin: United States

Manufacturer: Heath Kit

Height: 19½ in (50 cm)

Power source: Battery-operated

Intelligence: Microprocessor, self-diagnostics

Capabilities: Moves, speaks, detects movement, sounds an alarm

Self-Navigating Kit Robots

These clever kit robots are able to roam freely around their environments without instructions from an operator. By using light sensors and simple memory functions, they can navigate their way around by following the contours of a room or tracking dark lines marked on the floor.

Wall-Hugging Mouse

This Tamiya kit stops short of eating cheese, but in other ways it behaves like the real animal. It can track walls and change direction using a wire whisker that operates a micro switch along with a mechanical servo. The transparent body is reminiscent of a computer mouse but, like a real mouse, this one sports ears and a tail.

Specification: Wall-Hugging Mouse

First manufactured: 1983

Country of origin: Japan

Manufacturer: Tamiya

Height: 5¼ in (13.5 cm)

Power source: Battery-operated

Intelligence: Micro switch and mechanical servo operated by wall-hugging wire

Capabilities: Crawls, tracks wall

Hyper Line Tracker

This industrial-looking robot is made from an OWI-MOVIT kit. It has a memory system based on infrared sensors and phototransistors that allow it to track its way along a black line on a pair of motor-driven wheels. Despite its heavyweight design, Hyper Line Tracker has quite a simple mechanism illuminated by red LEDs that flash to indicate the robot's direction. The line can be made with a black marker or tape.

Wire carries power from the battery to the motor

Specification: Hyper Line Tracker

First manufactured: 1990

Country of origin: Japan

Manufacturer: MOVIT

Height: 4¼ in (11 cm)

Power source: Battery-operated

Intelligence: Light-sensor circuitry

Capabilities: Tracks a line

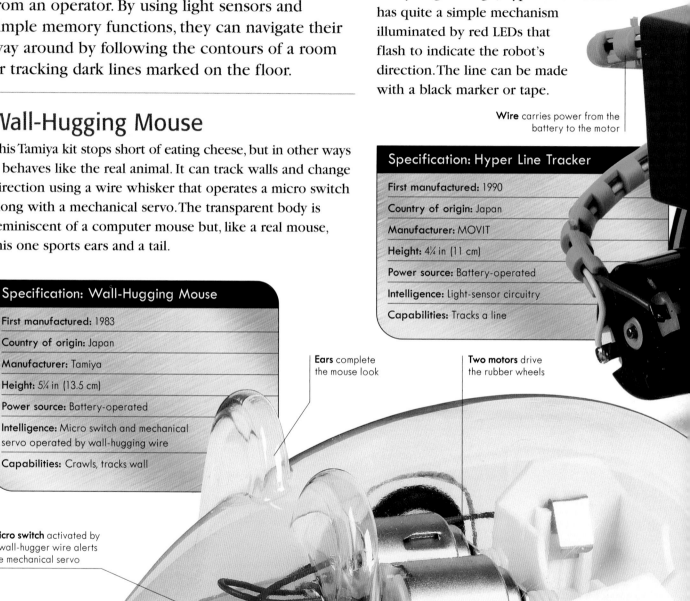

Ears complete the mouse look

Two motors drive the rubber wheels

Micro switch activated by a wall-hugger wire alerts the mechanical servo

Whisker is used to track the wall contours

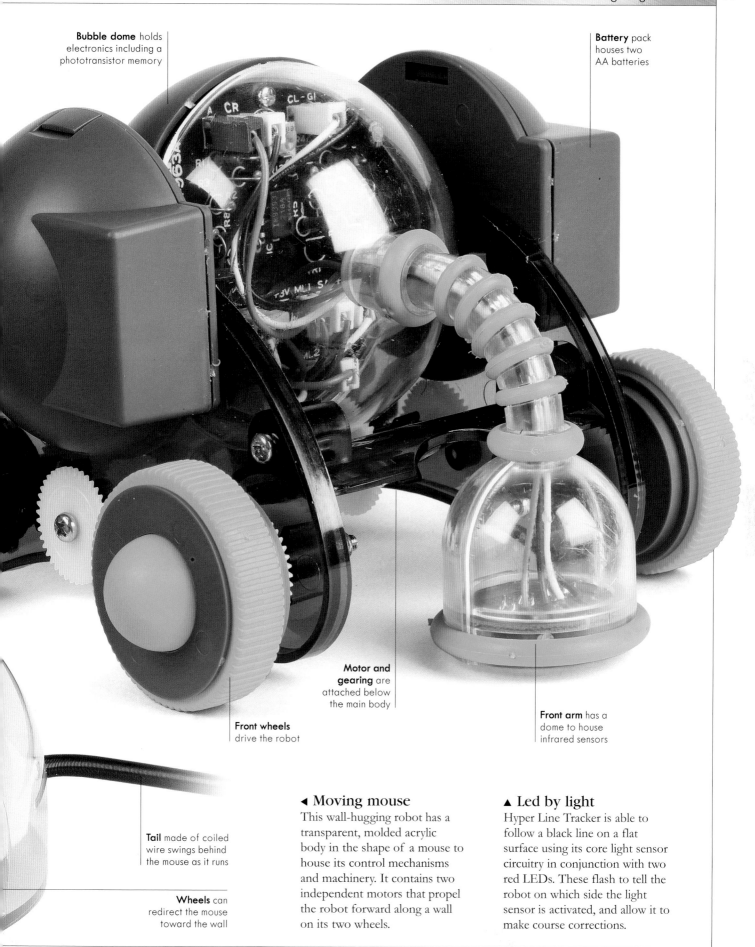

Bubble dome holds electronics including a phototransistor memory

Battery pack houses two AA batteries

Motor and gearing are attached below the main body

Front wheels drive the robot

Front arm has a dome to house infrared sensors

Tail made of coiled wire swings behind the mouse as it runs

Wheels can redirect the mouse toward the wall

◀ Moving mouse

This wall-hugging robot has a transparent, molded acrylic body in the shape of a mouse to house its control mechanisms and machinery. It contains two independent motors that propel the robot forward along a wall on its two wheels.

▲ Led by light

Hyper Line Tracker is able to follow a black line on a flat surface using its core light sensor circuitry in conjunction with two red LEDs. These flash to tell the robot on which side the light sensor is activated, and allow it to make course corrections.

Simple Kit Robots

Assembled from supplied parts that usually do not need soldering or any knowledge of electronics, simple kit robots are fun to build and play with. Many of the robots respond to voice, light, proximity, and touch. These entry-level kits use the same basic technologies as the industrial and humanoid robot world.

Specification: Medusa	
First manufactured: 1988	
Country of origin: Japan	
Manufacturer: OWI Inc.	
Height: 5¼ in (13.5 cm)	
Power source: Battery-operated	
Intelligence: Sensors and printed circuits	
Capabilities: Walks, responds to sounds	

Medusa

Named after the Gorgon of Greek mythology, but resembling the Martian spacecraft in H.G. Wells' novel *War of the Worlds*, this robot consists of a plastic dome and legs that bolt together, and comes with electronic parts preassembled. Activated by sound, such as a clap, its four-legged walking system uses rods to drive it forward.

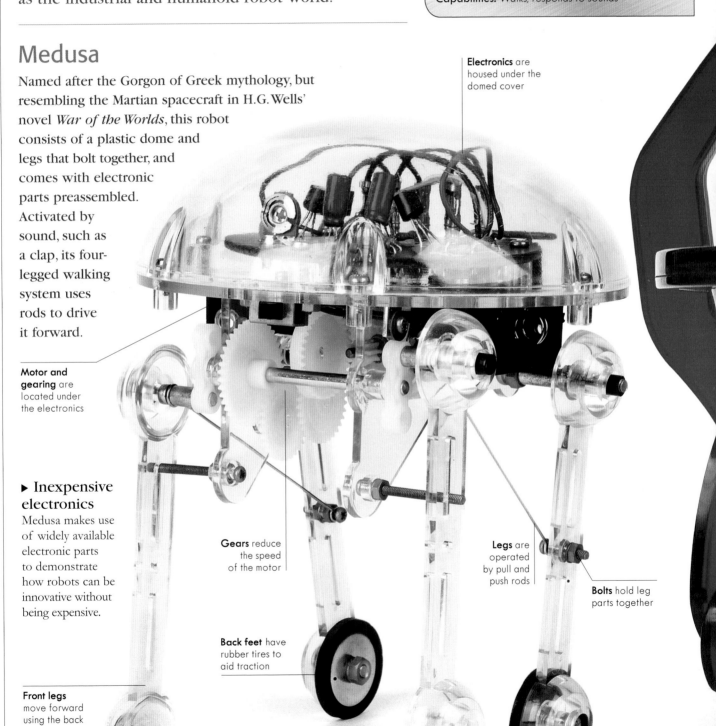

Electronics are housed under the domed cover

Motor and gearing are located under the electronics

▶ **Inexpensive electronics**
Medusa makes use of widely available electronic parts to demonstrate how robots can be innovative without being expensive.

Gears reduce the speed of the motor

Legs are operated by pull and push rods

Bolts hold leg parts together

Back feet have rubber tires to aid traction

Front legs move forward using the back legs as anchors

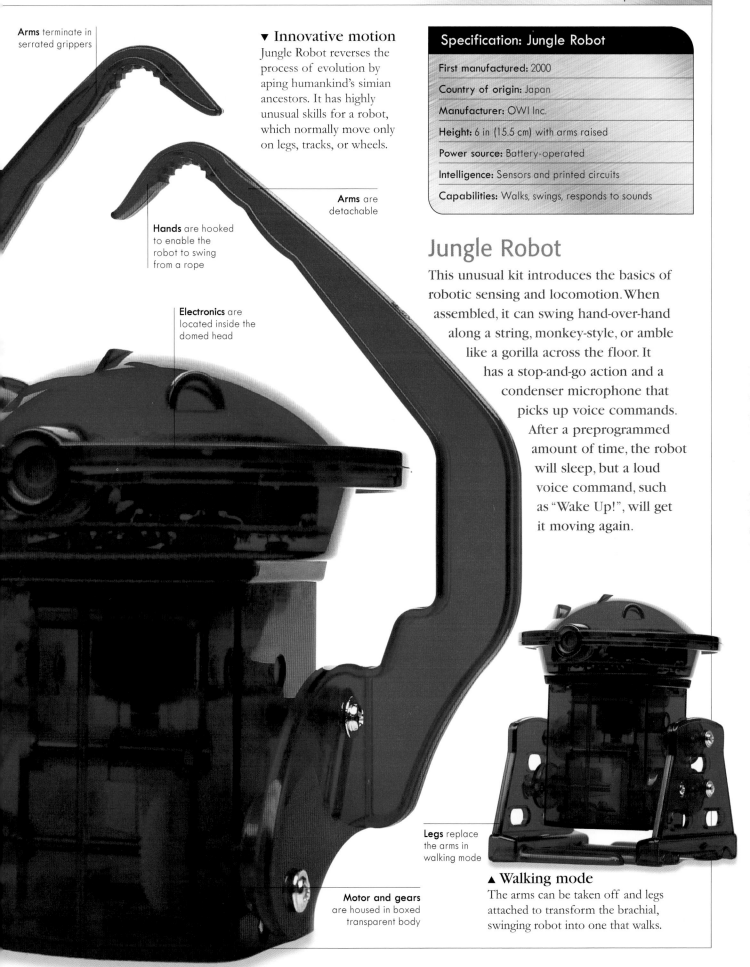

Arms terminate in serrated grippers

▼ **Innovative motion**
Jungle Robot reverses the process of evolution by aping humankind's simian ancestors. It has highly unusual skills for a robot, which normally move only on legs, tracks, or wheels.

Arms are detachable

Hands are hooked to enable the robot to swing from a rope

Electronics are located inside the domed head

Motor and gears are housed in boxed transparent body

Specification: Jungle Robot	
First manufactured: 2000	
Country of origin: Japan	
Manufacturer: OWI Inc.	
Height: 6 in (15.5 cm) with arms raised	
Power source: Battery-operated	
Intelligence: Sensors and printed circuits	
Capabilities: Walks, swings, responds to sounds	

Jungle Robot

This unusual kit introduces the basics of robotic sensing and locomotion. When assembled, it can swing hand-over-hand along a string, monkey-style, or amble like a gorilla across the floor. It has a stop-and-go action and a condenser microphone that picks up voice commands. After a preprogrammed amount of time, the robot will sleep, but a loud voice command, such as "Wake Up!", will get it moving again.

Legs replace the arms in walking mode

▲ **Walking mode**
The arms can be taken off and legs attached to transform the brachial, swinging robot into one that walks.

Creepy-crawlies

MINIATURIZATION IS CENTRAL to technology's progress, so it is no surprise that robots have begun to crawl, fly, and scurry like the smallest biological creatures. Today, there is a host of tiny robots that blend precision mechanics with the strange aesthetic appeal of insects. The minute scale of electronics has also brought bug-building into the reach of ordinary enthusiasts, with inexpensive robot kits.

▲ **Photopopper**
This robot can cover 3 ft 4 in (1 m) in two minutes. It is solar-powered and has infrared detectors for positioning and obstacle avoidance.

Clockwork bugs were old news by the nineteenth century: the triumphs of the watchmaker's art, which excelled in building intricate mechanisms on a miniature scale. Today, our fascination with technology still goes hand in hand with our delight in miniature objects. Smaller means better, so what better way to show our mastery of technology than by building micromachines?

Small is beautiful

In the past, it was possible to produce robot toys that could imitate a walking action only rather crudely with wind-up mechanisms. Microprocessors, printed circuitry, miniature sensors, and tiny motors have changed all that, and the minute scale of today's electronics makes bug-building with inexpensive kits a hobby that can be enjoyed by anyone. The parts may include motors the size of a pea and components no larger than a grain of rice. Some robot bugs even carry tiny digital cameras.

In the real world, bugs can intrigue as well as repel us—much like our relationship with new technology. When viewed microscopically, real bugs appear strangely beautiful, with complex eyes and intricate bodies, wings, and limbs. Robot bugs are almost equally fascinating, with their displays of microcircuitry, sensors, and delicate antennae.

Some robots, such as many BEAM (Biology Electronics Aesthetics and Mechanics) kits, accomplish their movement via solar power arrays positioned on their backs. Others use traditional battery-powered motors.

▲ **Hydrazoid**
This long-legged bug reacts to sound and uses the signal to move forward for 15 seconds.

Large family

In the robot bug world, there are thousands of hybrid species, from two-legged scarabs and four-legged spiders to giant fleas and friendly roaches. Most are more interactive than their real-world counterparts,

Six-legged system
gives stability and speed

▲ **Antoid**
Able to detect objects up to 20 in (50 cm) away with its infrared eye, the Antoid walks on six paired legs.

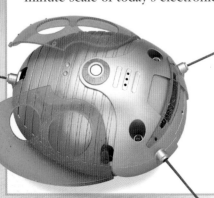

◄ **IQbug**
This singing robot reacts to sound and light and its eyes flash to express emotion.

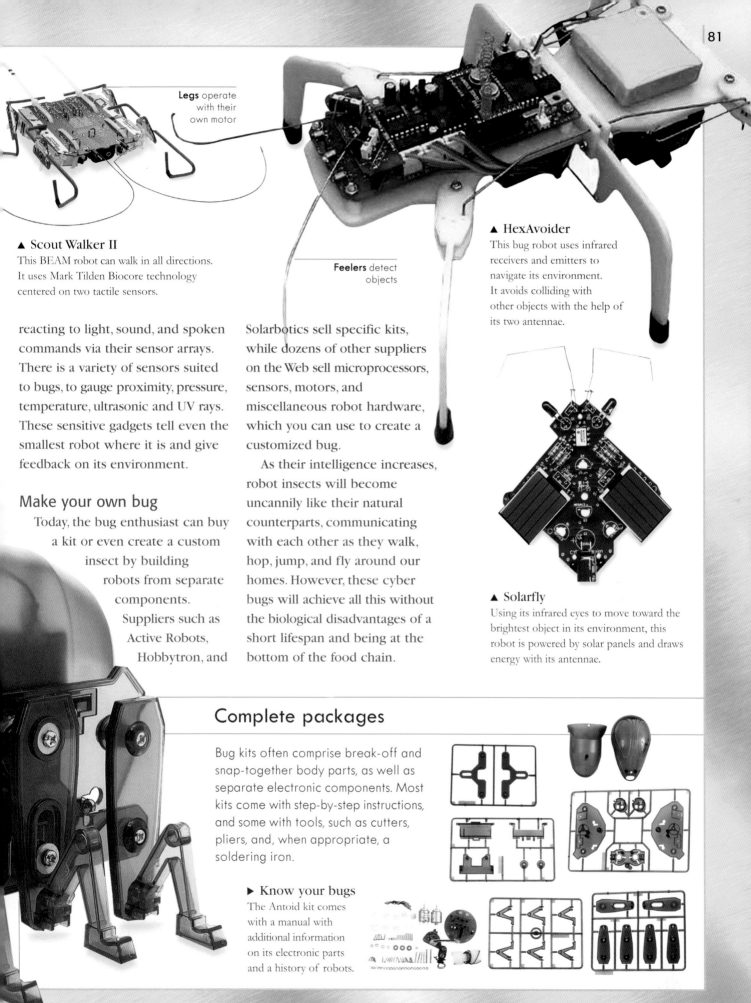

Legs operate with their own motor

Feelers detect objects

▲ Scout Walker II
This BEAM robot can walk in all directions. It uses Mark Tilden Biocore technology centered on two tactile sensors.

▲ HexAvoider
This bug robot uses infrared receivers and emitters to navigate its environment. It avoids colliding with other objects with the help of its two antennae.

reacting to light, sound, and spoken commands via their sensor arrays. There is a variety of sensors suited to bugs, to gauge proximity, pressure, temperature, ultrasonic and UV rays. These sensitive gadgets tell even the smallest robot where it is and give feedback on its environment.

Make your own bug
Today, the bug enthusiast can buy a kit or even create a custom insect by building robots from separate components. Suppliers such as Active Robots, Hobbytron, and

Solarbotics sell specific kits, while dozens of other suppliers on the Web sell microprocessors, sensors, motors, and miscellaneous robot hardware, which you can use to create a customized bug.

As their intelligence increases, robot insects will become uncannily like their natural counterparts, communicating with each other as they walk, hop, jump, and fly around our homes. However, these cyber bugs will achieve all this without the biological disadvantages of a short lifespan and being at the bottom of the food chain.

▲ Solarfly
Using its infrared eyes to move toward the brightest object in its environment, this robot is powered by solar panels and draws energy with its antennae.

Complete packages

Bug kits often comprise break-off and snap-together body parts, as well as separate electronic components. Most kits come with step-by-step instructions, and some with tools, such as cutters, pliers, and, when appropriate, a soldering iron.

▶ Know your bugs
The Antoid kit comes with a manual with additional information on its electronic parts and a history of robots.

Cybot

A spinoff project of the Department of Cybernetics at Reading University in the UK, Cybot is inspired by the Seven Dwarves project, which studied how robots could act in groups. Each issue of *Ultimate Real Robots* magazine offers part of the Cybot kit, comprising the robot, and the remote-control headset.

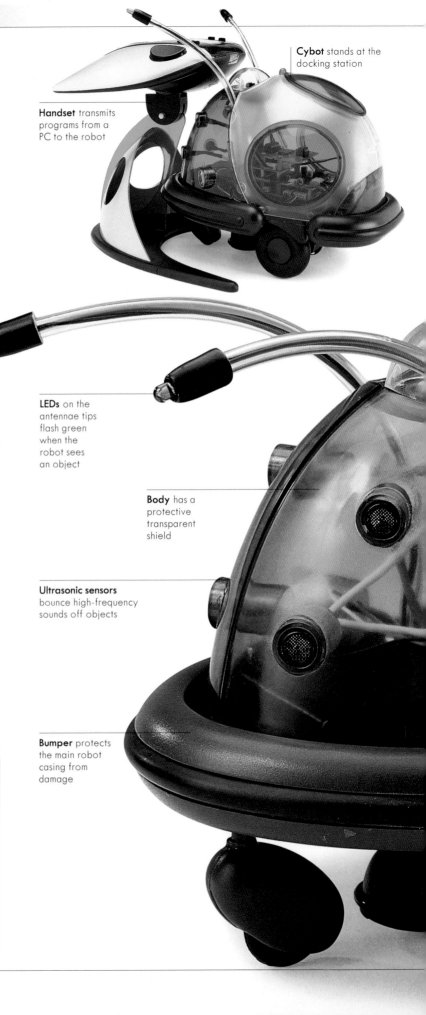

Cybot stands at the docking station

Handset transmits programs from a PC to the robot

Sophisticated sensing robot

Cybot has light and ultrasonic sensors, which enable it to seek or avoid light sources, avoid solid objects, follow moving objects, and follow lines. Its motors, powered by AA batteries, use a set of gears to increase torque to the back wheels. The front caster wheel balances the robot and allows sharp turns. Cybot is controlled by up to seven microprocessors, plus three in the handset.

LEDs on the antennae tips flash green when the robot sees an object

Body has a protective transparent shield

Ultrasonic sensors bounce high-frequency sounds off objects

▶ **Elegant exterior**

Fully assembled, stylish Cybot makes a splash in its cool colors. Four ultrasonic sensors, which bounce sound waves off objects to gauge their distance protrude through the body shield. The front wheel rotates through 360 degrees, allowing Cybot to turn within its own length, or spin in place

Bumper protects the main robot casing from damage

Specification: Cybot

First manufactured:	2001
Country of origin:	UK
Manufacturer:	Licensed to Eaglemoss Publications
Height:	7 in (17 cm)
Power source:	Battery operated
Intelligence:	Microprocessors in robot and handset
Capabilities:	Seeks and avoids light, avoids and follows objects, follows lines, recognizes voice commands, plays ball

◄ **PC-friendly**

Cybot can be driven and put through its various modes via a remote control infrared handset. Supported on its docking station (pictured) and connected to a PC, programs can be downloaded to the robot. A microphone headset also plugs into the handset to enable Cybot to respond to the user's voice commands.

Dome houses transceiver for communication with handset

Flippers enable the robot to handle the ball

Goal contains an infrared beacon to guide Cybot

◄ **Playing Cyball**

With the addition of a motorized flipper, two arms, and a roller-ball caster, the robot can play a ball game called Cyball. Sensors in the robot's arms track infrared beacons inside both a special ball and a goal tag. The robot can find the ball, guide it toward the goal, and flip the ball forward to score—all by itself or with other Cybots.

Ball has an infrared tag so that the robot can detect it

Panels allow access to robot's mode selector switch

▼ **On line**

Cybot can detect and follow the edge of a black or dark-colored line against a pale surface. It does this by utilizing two infrared sensors which feed near-ground inputs into a comparator; this, in turn, feeds the main processor.

Infrared sensors for line tracking

Wheels at the rear are powered by electric motors

LEGO Spybotics

In these imaginative robots, LEGO combines traditional model-making with computer technology. Each kit makes a single, sensor-packed robot with a different combination of skills, speed, strength, and agility. Once the robot is built and programmed, there are downloadable spy missions that add an extra dimension to play, teaching the owners about remote-controlled robotics while involving them in a fantasy world of espionage.

Spybotics secret agents

When someone buy a Spybotics kit, they "become" an agent working for the ultimate spy agency, S.M.A.R.T, via the Spybotics website and CD-ROM. To complete missions, they will work with a specially designed robot partner, known as a Spybot, which they build from a kit. Each of these robots is designed to help complete set tasks that match its particular strengths. Owners need a serial cable and a PC to use the tools, missions, games, and bonus skills.

Specification: Snaptrax	
First manufactured: 2003	
Country of origin: Denmark	
Manufacturers: LEGO Group Inc.	
Height: 4¾ in (11.4 cm)	
Power source: Battery-operated	
Intelligence: PC interface	
Capabilities: Has tractor locomotion for agility and strength, uses beetle-style pincers to grab objects	

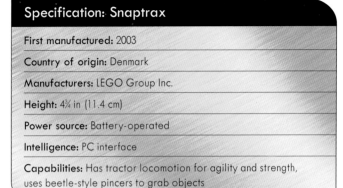

Treads add traction to mobility

▲ **Snaptrax**
Scoring high on agility and strength, this bulldozer-style robot is created from 193 different pieces, including pincers that snap at anything in its path.

Pincers can be programmed to snap together

Specification: Gigamesh	
First manufactured: 2003	
Country of origin: Denmark	
Manufacturer: LEGO Groups Inc.	
Height: 4¾ in (11.4 cm)	
Power source: Battery-operated	
Intelligence: PC interface	
Capabilities: Has strength, moves spiderlike over rough terrain, clears path with upfront grinder	

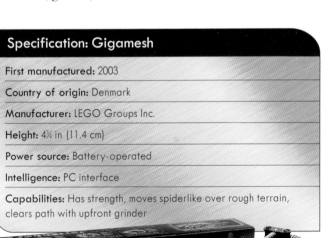

Wheels have deep treads for extra traction

Light flashes as the robot moves

Wheels have gears for all-terrain locomotion

◀ **Gigamesh**
This robot's advantage is its strength. Close to the ground and wide, its many wheels make it very stable. The kit comes in 233 pieces, and a programmable controller allows you to customize it for a mission.

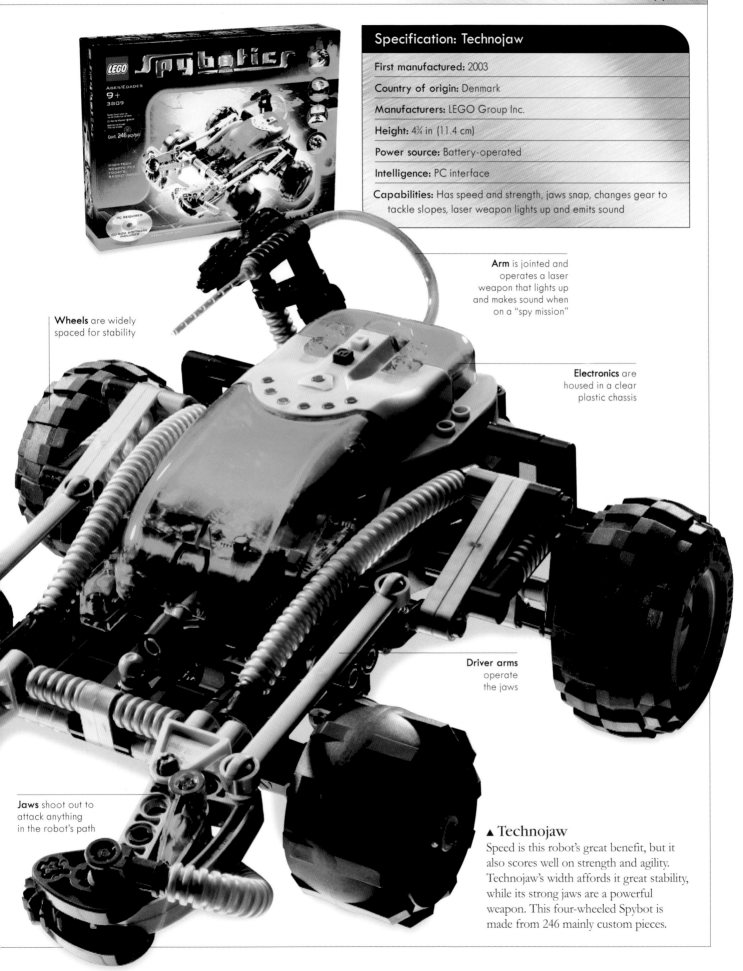

Specification: Technojaw

First manufactured: 2003

Country of origin: Denmark

Manufacturers: LEGO Group Inc.

Height: 4¾ in (11.4 cm)

Power source: Battery-operated

Intelligence: PC interface

Capabilities: Has speed and strength, jaws snap, changes gear to tackle slopes, laser weapon lights up and emits sound

Arm is jointed and operates a laser weapon that lights up and makes sound when on a "spy mission"

Wheels are widely spaced for stability

Electronics are housed in a clear plastic chassis

Driver arms operate the jaws

Jaws shoot out to attack anything in the robot's path

▲ Technojaw

Speed is this robot's great benefit, but it also scores well on strength and agility. Technojaw's width affords it great stability, while its strong jaws are a powerful weapon. This four-wheeled Spybot is made from 246 mainly custom pieces.

LEGO MindStorms

Many of us have grown up with LEGO building bricks, but today's LEGO also provides kits that teach users about design, technology, computing, and robotics. LEGO bricks still make up the body, but microcomputers, sensors, communication devices, and software are added to make imaginative robots.

Specification: MindStorms
First manufactured: 2004
Country of origin: Denmark
Manufacturer: LEGO Group, Inc.
Height: Various
Power source: Battery-operated
Intelligence: RCX microcomputer
Capabilities: Walks, runs, senses objects

Robotics Invention System

As with most LEGO projects, the company provides the bricks, but the builder is free to create whatever fires the imagination. A MindStorms starter kit comprises 717 parts, including motors, wheels, gears, software, an RCX microcomputer, touch sensors, a light sensor, and an infrared transmitter. Some of these devices are placed within a conventional LEGO brick. The manufacturer suggests that 25 different robots can be built from the basic kit.

▲ **Robotics building bricks**
This upgraded kit includes more powerful software and additional parts that allow you to build a variety of robots that follow simple commands. An Ultimate Builders set is available for the advanced kit builder.

◀ ▼ **Varied forms**
These two robots show how versatile the kits are. One is a simple walker while the other is a fully equipped and geared rolling robot.

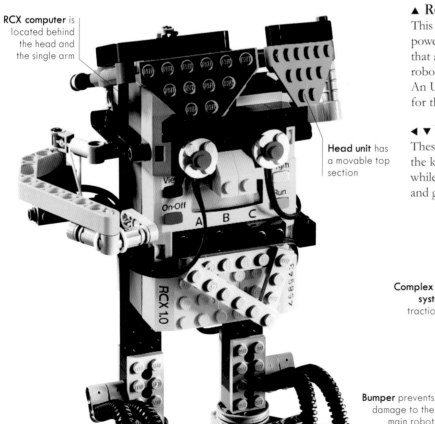

RCX computer is located behind the head and the single arm

Head unit has a movable top section

Feet can move up and down

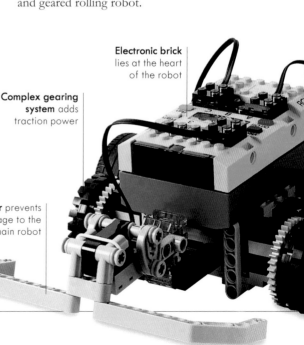

Electronic brick lies at the heart of the robot

Complex gearing system adds traction power

Bumper prevents damage to the main robot

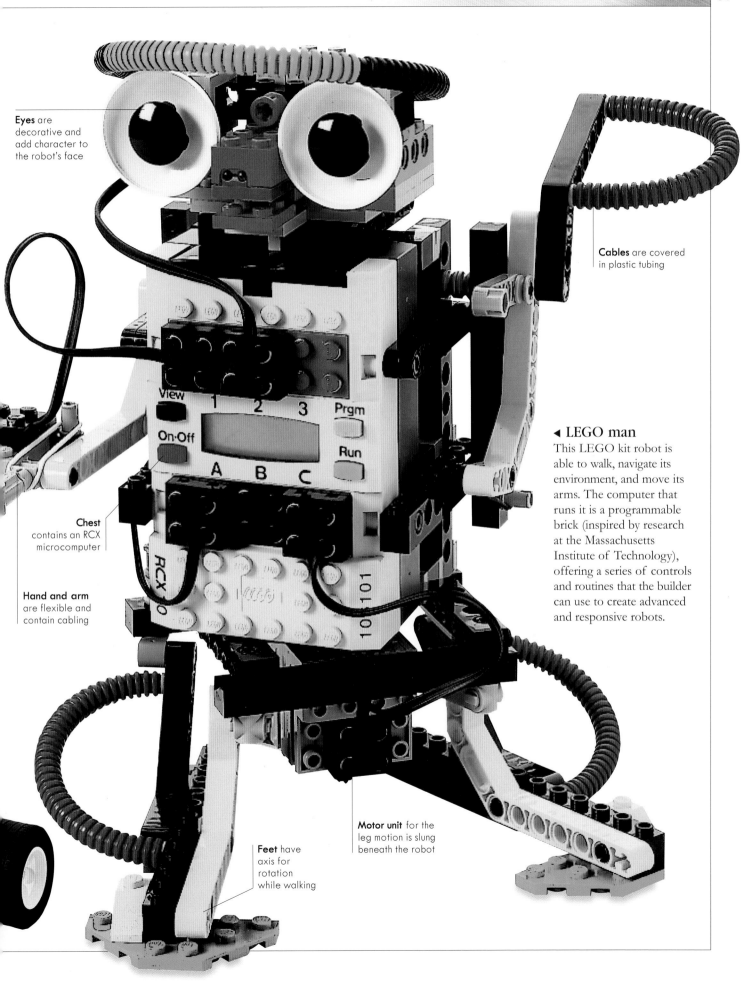

Eyes are decorative and add character to the robot's face

Cables are covered in plastic tubing

View 1 2 3 Prgm
On·Off A B C Run

RCX

10 101

Chest contains an RCX microcomputer

Hand and arm are flexible and contain cabling

◄ **LEGO man**
This LEGO kit robot is able to walk, navigate its environment, and move its arms. The computer that runs it is a programmable brick (inspired by research at the Massachusetts Institute of Technology), offering a series of controls and routines that the builder can use to create advanced and responsive robots.

Feet have axis for rotation while walking

Motor unit for the leg motion is slung beneath the robot

Humanoid Kits

CAM and Dr. Robot kits are humanoids at opposite ends of the robot evolutionary scale, but all show us how robots are learning to be human. The CAM robots' walking style is giant leaps ahead of the small steps of early toys, while Dr. Robot's web-surfing HR6 rivals the most advanced robots made.

CAM-08

Taking its name from its internal cam-driven mechanism, this simple kit works by mechanics rather than computer technology. It has flexible foot and knee joints, and walks, arms swinging, with a natural gait.

Legs move by cam technology

Head clips on to body and rotates by motor action

Abdomen houses a small internal speaker

▶ **Human stride**

As it walks, CAM-08's head turns left and right and the robot emits an electronic sound. The robot's arms swing in rhythm as it moves. The batteries are housed in the robot's backpack, and the robot is remote-controlled.

Specification: CAM-08

First manufactured: 2003

Country of origin: Japan

Manufacturer: Cube Co. Ltd

Height: 8 in (20 cm)

Power source: Battery-operated, remote control

Intelligence: None

Capabilities: Walks backward and forward, swings arms, turns head, makes sounds

CAM-10

The smaller CAM-10 comes preassembled, but it can be taken apart to display its inner workings. It is more advanced than CAM-08, thanks to a more complex mechanism of gears and cams.

Head turns as robot walks

Chest panel can be removed to change the batteries

Knee joint is activated by an eight-piece cam

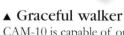

▲ **Graceful walker**

CAM-10 is capable of only forward motion, but steps on its jointed feet with real humanoid grace. It walks with a purring motor and gear noise, while its arms swing back and forth and its head turns.

▶ **Choice of colors**

The robot is available in three different colors: pink, white, and transparent blue—which displays the internal mechanism.

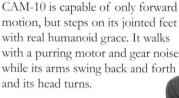

Specification: CAM-10

First manufactured: 2004

Country of origin: Japan

Manufacturer: Cube Co. Ltd.

Height: 4 in (10 cm)

Power source: Battery-operated

Intelligence: None

Capabilities: Walks forward, arms swing, head turns

Dr. Robot HR6

This top-of-the-line robot is designed for robotic research at universities. It has a wireless module, PC software, and an optional voice-recognition system. It can walk, dance, sit, recognize faces, respond to voices, and even surf the Internet. When its battery runs low, it returns to its recharging station.

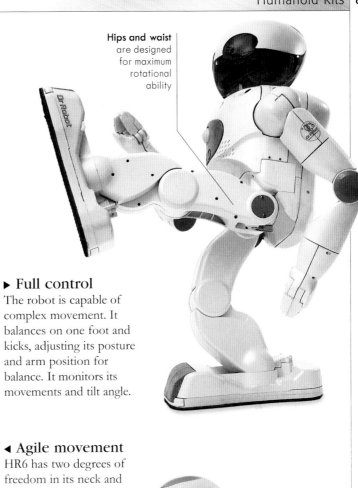

Hips and waist are designed for maximum rotational ability

Head contains a color camera

Chest contains stereo and audio output and a microphone

Legs have five rotational joints

Large feet aid balance during complex movements

▶ Full control
The robot is capable of complex movement. It balances on one foot and kicks, adjusting its posture and arm position for balance. It monitors its movements and tilt angle.

◀ Agile movement
HR6 has two degrees of freedom in its neck and head, in addition to its 20 degrees of freedom in the arms and legs. A motor governs every synchronized joint.

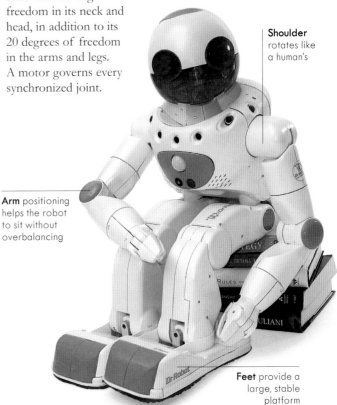

Shoulder rotates like a human's

Arm positioning helps the robot to sit without overbalancing

Feet provide a large, stable platform

▲ Sitting down
Only a robot with fully synchronized motion can actually sit down—a skill even humans have to learn. It has to bend forward and fall backward, but not over.

Specification: Dr. Robot HR6

First manufactured: 2003

Country of origin: Canada

Manufacturer: Dr. Robot

Height: 20½ in (52 cm)

Power source: Battery-operated

Intelligence: Wireless PC and Internet connection

Capabilities: Walks, sees, sits, lies down, dances, recognizes faces, gets up from prone position

GALLERY: Robot Construction Kits

From simple toys to advanced humanoids, robot construction kits cater to every level of interest and ability. Some are built purely for fun; others are intended for educational use. Some are designed to compete in robot contests, while others are built with shared knowledge from the Internet. Most are user-friendly DIY packages bought off the shelf, and the components can include motors, sensors, lights, computer controllers, and software. The finished robot may resemble a human, a cyber insect or animal, a Space Age vehicle, or a creature from outer space.

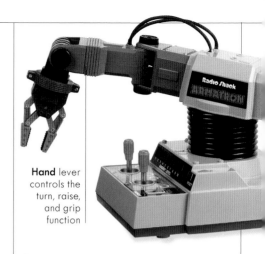

Hand lever controls the turn, raise, and grip function

Armatron

This battery-run toy demonstrates robot arm motion and gripping control.
■ Date: 1978 (approx) ■ Country of origin: United States/Japan ■ Height: 7½ in (18.3 cm)
■ Manufacturer: Radio Shack/Tomy

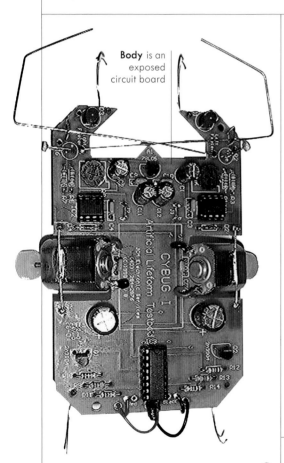

Body is an exposed circuit board

Cybug scarab

This insectlike motor-driven robot can avoid obstacles and walk backward. It is programmed to be most active at night, and can seek or avoid light.
■ Date: 1998 ■ Country of origin: Japan ■ Height: 11½ in (29.2 cm)
■ Manufacturer: OWI

Walking Owl

This friendly-looking owl is an entry-level kit. It teaches the user to control its movements via four RC servo motors that are run by a program downloaded to the onboard CPU.
■ **Date:** 2002
■ **Country of origin:** UK
■ **Height:** 4½ in (11.5 cm)
■ **Manufacturer:** Active Robots

Legs are powered by computer-driven motors

Body houses infrared and light sensors used to change direction

Wonderborg

This self-controlled robot with six legs can be programmed by a PC. It changes direction when it encounters an obstacle.
■ Date: 2000 ■ Country of origin: Japan
■ Height: 3¼ in (8 cm) ■ Manufacturer: Bandai

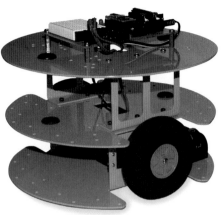

Rogue Blue Simmbot

This learning tool has preprogrammed instructions and can accept modular upgrades.

- **Date:** 2002 ■ **Country of origin:** Canada
- **Height:** 6 in (15 cm) ■ **Manufacturer:** Rogue Robotics

Linebotz

This kit robot moves on two wheels using directional built-in sensors to follow a line.

- **Date:** 2003
- **Country of origin:** UK
- **Height:** 4 in (10 cm) approx
- **Manufacturer:** Ibotz

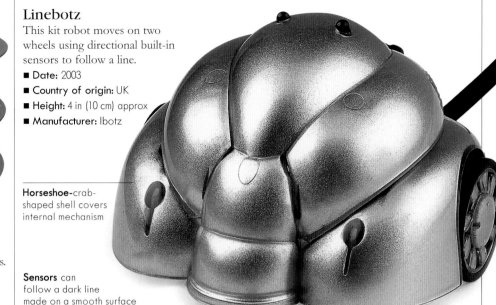

Horseshoe-crab-shaped shell covers internal mechanism

Sensors can follow a dark line made on a smooth surface

Airat 2

A kit made for the Micro Mouse contests, Airat 2 uses six sensors and a searching algorithm to find its way around a maze.

- **Date:** 2002
- **Country of origin:** UK
- **Height:** 4½ in (11.4 cm)
- **Manufacturer:** Active Robots

Soundtracker

This easy-to-assemble plastic three-wheeled kit has sound sensors with a 20-in (50-cm) range. It reverses away from objects and then moves forward.

- **Date:** 2003 ■ **Country of origin:** UK ■ **Height:** 3¾ in (9.5 cm)
- **Manufacturer:** Ibotz

Electronics are protected by the plastic shell

Tires are ribbed to aid traction

Snout houses a sensor

Wheels reverse when sensor picks up a high-pitched sound or vibration

Cyber Raptor

Part of a range of combat vehicles, Cyber Raptor can be assembled by enthusiasts to take part in simulated battles in theme arenas.

- **Date:** 2003
- **Country of origin:** UK
- **Height:** 36cm (14½in)
- **Manufacturer:** RobotsRus

Casing is designed to withstand combat

Four-wheel drive

Flipper weapon is attached to the front of the robot

Robots as Cultural Icons

SCIENCE FICTION describes imaginary futures, but it also speaks volumes about the era in which it is created. Robots are iconic partly because they reflect the design aesthetics and technology of their period. However, they also reflect social and cultural trends. As machines made in our own image, robots allow us to explore what it means to be human whenever human life seems threatened.

A scene from the 1954 movie *Tobor the Great*, in which a robot astronaut is controlled by a boy

Although they are machines, robots have often been used to portray the human preoccupations of the times. In the 1920s and 1930s, robots were used by writers and artists to reflect our fears about totalitarian states and about how technology might turn us into machines. At the height of Cold War paranoia about Communism, alien robots were depicted as a threat to the nuclear family and the American Dream. In the 1960s and 1970s, the era of both the hippie movement and the rise of computers, robots were used to explore the clash between emotionless machines and the natural world. In the consumer-driven 1980s, robots were often portrayed as the products of huge, faceless corporations, and

A 1950s board game in which a robot pointed to 150 quiz answers

in the 1990s, anxiety about the approaching millennium saw robots portrayed as human beings' all-powerful replacements. Most recently, robots have featured in stories about how artificial life might develop alongside human society, at a time when technology is accepted but genetic research and human rights are major issues.

The human versus the machine

Robot stories, films, and art explore what it means to be human, asking: what would we be like without emotions, or with superhuman strength? Or, would being immortal or infallible make us better people? Karel Capek's original robots, in his 1921 play *R.U.R.*, were human beings turned into machines by

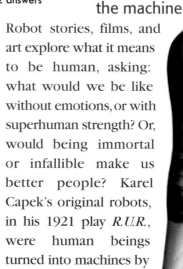

Andrew in *Bicentennial Man* (1999), who wants to be recognized as a human being

Cover illustrations for the 1976 Dalek annual (below), and the 1954 novel *The Humanoids* (right)

A 1970s advertisement for instant mashed potatoes (left), and a 2003 advertisement for beer (right)

Robot clowns, heroes, and villains

Where robots tread, humor is rarely far behind—perhaps because they so closely resemble us. Advertisers have used drunken robots to promote alcoholic drinks, and laughing alien robots to sell us instant mashed potatoes. Robot iconography has been explored by many artists across a range of media, including painting, sculpture, and animation. While many are attracted to the comical side of robots, using them to satirize human foibles and weakness, others are inspired by their aesthetics—whether sleek, industrial, or whimsically nostalgic. Indeed, the pulp-fiction illustrations and science-fiction jacket designs of the 1930s to the 1970s have done more to influence our concept of robots

Wrestling robots at the Robo-One Grand Prix

than even movies have done. Artists turned writers' ideas into visions of amazing worlds and warlike, alien robots destroying humanity. Today, the amazing graphics of computer games give us the illusion of having robot powers ourselves. Robots are even making an impact in the world of sports, with soccer tournaments taking place all over the world. From Capek to Spielberg, robots have become icons of the art and entertainment world.

repetitive jobs and the rise of mass production. Maria, the evil robot in Fritz Lang's 1926 movie *Metropolis*, was made in the image of a compassionate woman. In contrast, the first robot celebrity, Robby, was clearly a machine, but he was also a loyal companion to his human master.

Robots that feel

This was Isaac Asimov's vision: robots serving and augmenting human beings with machine intelligence and strength. In his story *Bicentennial Man* (made into a movie in 1999), a robot tries to be human for 200 years, eventually choosing to die alongside his human partner. To be human, he learns, is to be mortal, and to love.

A scene from the 2001 movie *A.I.*, showing David, the "mecha" robot boy, and his robot teddy bear

Some of the most famous fictional robots explore the differences between humans and machines. Steven Spielberg's movie *A.I.* (2001), based on a story by Brian Aldiss, features a robot who wants to be a real boy so that he can experience love. The robots of the *RoboCop* and *Terminator* films find greater strength through human loyalty and self-sacrifice. In *Silent Running* (1971), Huey, Dewey, and Louie's devotion to the ecosystem makes them more human than the people who want to destroy it. In *The Hitchhiker's Guide to the Galaxy*, Marvin's brain is so vast that he experiences a superhuman depression.

A scene from the *Transformers* TV series, which ran for 98 episodes from 1984

Maria

Cinema's first robot icon was female. Fritz Lang's Maria stands at the heart of the great silent film *Metropolis*. Like the city, she is a stunning creation on the surface, but is dark and inhuman beneath. She is an evil replica of a woman who is honest and true.

Metropolis

A huge city reminiscent of modern-day New York or Tokyo, Metropolis is made up of two social strata: a rich ruling class and an enslaved class who live in a hellish underground power plant. Rotwang, the archetypal mad scientist, replaces Maria— a compassionate woman who befriends the workers—with a robot duplicate that brings the city to the verge of destruction.

Movie poster, 1926

◄ **Power-mad**
Maria is prepared for her mission. Rotwang captures the real Maria, and copies her face and body onto the robot's metal surface, making her indistinguishable from the real woman. The scene is set for the robot to bring chaos and rebellion to the ordered and peaceful surface of the great city.

▶ **Silent presence**
Rotwang (right), played by Rudolph Klein-Rogge, rages at Ferdersen, the ruler of Metropolis, played by Alfred Abel. Silent films were characterized by over-dramatic acting, so the understated movements of the robot contributed to her iconic status.

Head is fully
armored, with
gaps for the eyes
and mouth only

Neck has a
capelike form
that accentuates
the shoulder curves

Arms are
designed to look
like mechanically
joined parts

Specification: Maria

Date of film debut:	1926
Country of origin:	Germany
Studio:	UFA Studio, Berlin
Height:	5 ft 6 in (1.7 m)
Capabilities:	Walks, dances, impersonates a human

◀ Timeless icon

This 1985 vinyl model of the Maria robot, made by
Matsudaya, stands 17½ in (44 cm) high. It is one of
many movie-related figures flooding the collectors'
market today. Long after Fritz Lang's original
film was made, Maria is still a cult figure and an
undoubted classic of science-fiction costume design.

▶ Humanoid design

The false Maria is a beautiful Art Deco–styled robot.
Although she was modeled on the actress playing
Maria, Brigitte Helm, the designers added machinelike surface details to create the first robot with a
truly human figure. The design inspired many later
film robots, including C-3PO in the *Star Wars* movies.

Legs emphasize
the Art Deco–style
sculptural design

Shins and feet
suggest medieval
armor

Gort

A huge interstellar guardian, Gort was one of the first robot movie stars. He is empowered by an alien race to destroy worlds and prevent hostile, aggressive beings, such as humans, from spreading throughout the galaxy.

The Day the Earth Stood Still

Made at the height of the Cold War, this movie, directed by Robert Wise, has a message: stop the spread of atomic weapons or Earth will be destroyed. The compassionate alien Klaatu, accompanied by the silent law enforcer Gort, wants to meet the world's philosophers, only to be faced by its armed forces. To demonstrate his power, Gort shuts down the world's power supplies as a warning to humankind—giving the movie its title.

FROM OUT OF SPACE...
A WARNING AND AN ULTIMATUM!

THE DAY THE EARTH STOOD STILL

MICHAEL RENNIE PATRICIA NEAL HUGH MARLOWE

Film poster, 1951

▶ **Metal giant**
This is a full-size replica of the 9-ft (2.7-m) prop used to represent Gort in the movie. An actor in an 8-ft-2-in (2.4-m) costume played the scenes in which the robot walked. Gort's design is echoed in *RoboCop* and the Pino range of toys.

Face is featureless

Head is similar to a police helmet

Visor slides upward for the robot to scan objects and unleash the death ray

Shoulders are broad to emphasize the robot's power

▶ **Collector's item**
This modern Gort figurine is made by Rocket. It is a 5-in (12.5-cm-) high cast-metal copy of the original prop, and is marketed as a collector's item rather than as a toy.

GORT

METAL COLLECTABLE

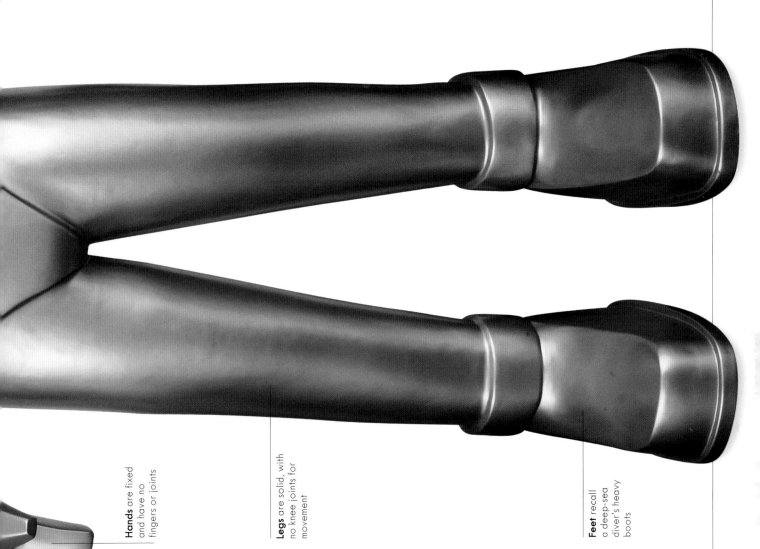

Hands are fixed and have no fingers or joints

Legs are solid, with no knee joints for movement

Feet recall a deep-sea diver's heavy boots

Specification: Gort

Date of film debut: 1951

Country of origin: United States

Studio: 20th Century Fox

Height: 9 ft (2.7 m)

Capabilities: Can destroy worlds with lethal heat ray, eliminates Earth weapons, defends Klaatu

▲ Interstellar visitors

Klaatu (played by Michael Rennie) steps out of his flying saucer after landing in Washington, DC, with his robot bodyguard Gort. Despite coming in peace, Klaatu is mistaken for an aggressor by Earth's armed forces. In the scenes where the robot had to walk, Gort was played by Lock Martin, who, at 7 ft 7 in (2.3 m) tall, was one of Hollywood's tallest actors.

Robby the Robot

Here is a science-fiction legend. Although he was not the first movie robot, Robby was the first robot movie star, and one of cinema's first merchandising opportunities. He appeared in two movies, and made guest appearances in several TV shows, including *Lost in Space*, alongside B9, the robot he had inspired. Robby was immortalized in a range of toys that capitalized on his appeal both as a friend and as a potentially dangerous adversary.

Head is dome-shaped and has flashing lights

Hands have powerful grip

Forbidden Planet

Robby debuted in *Forbidden Planet*, a movie directed by Fred McLeod Wilcox, which was inspired by William Shakespeare's play *The Tempest*. Robby is the devoted assistant to his maker, Dr. Morbius, on the planet Altair IV. Spaceship commander John J. Adams and his crew land on the planet to investigate the disappearance of an earlier mission, only to find Morbius, his daughter Altaira, and Robby. The visitors soon discover that an ancient race still influences the planet and a deadly force is terrorizing the inhabitants.

Movie poster, 1956

▲ **Reaching out**
One of the crew from the visiting spacecraft overcomes his initial fear of the alarming-looking Robby and discovers he is not, in fact, an evil machine, just a loyal servant to his creator, Dr. Morbius.

Body shape is based on a potbelly stove

◄ **Robot toys**
Robby spawned many toys, such as this 1984 robot reproduction by Matsudaya. The simple wind-up toy in plastic is shown with its original box. Many other robot toys have clearly been inspired by Robby, imitating aspects of his design, especially his bulbous legs, short arms, and domed head.

Specification: Robby the Robot
Date of film debut: 1956
Country of origin: United States
Studio: MGM
Height: 7 ft (2 m)
Capabilities: Walks, has superhuman strength, antenna rotates, lights flash, pistons pump

Antenna is motorized and rotates

Head mechanism suggests eyes, nose, and grinning mouth

◀ Actor in disguise

Although Robby was played by an actor, he bore only a vague resemblance to a human being. The movie's design team wanted audiences to think of him as a machine, and not as an actor in a costume. Robby was manned and voiced by Frankie Carpenter and Frankie Darro.

Arms have superhuman strength

Costume is made from lightweight vacuum-formed plastic

The Invisible Boy

Robby was the real star of *Forbidden Planet*, and its huge success meant that a second movie was inevitable. In 1957, he starred in the MGM follow-up, *The Invisible Boy*. A scientist's son, Timmy, discovers Robby (brought back from the future in a time-travel experiment) in his father's laboratory. They become friends, but Robby's built-in "conscience" bores the boy, and he changes the robot's program using Univac, his father's computer. However, Univac has been infected by an alien intelligence that wants to rule the world, and it takes control of the robot.

▲ Evil Robby

Robby's second appearance was an early portrayal of the threat to humans from computer technology, in particular from computer viruses, as we know them today.

Movie poster, 1957

▼ Dream of a generation

The bond between Robby and young Timmy symbolized the fascination with space and technology felt by children of the 1950s and 60s, who grew up during the height of the space race.

The Daleks

The race of mutant, pepperpot-shaped monstrosities from the planet Skaro has been terrifying children since their 1963 debut in the British TV series, *Doctor Who*. Under the command of their mad master, Davros, the Daleks patrol the galaxy—their mission: universal domination. With their fearsome mechanical cries of "Exterminate!" and "You will obey!", they set out to enslave humanity and all other species. Those they cannot subjugate they will destroy.

Specification: Dalek

Date of film debut: 1965	
Country of origin: UK	
Studio: Regal Films and Lion International	
Height: 5 ft (1.5 m)	
Capabilities: Head and upper body rotate, gun stick fires lethal death ray, solar panels convert energy	

▶ **Unusual inspiration**
The distinctive shape and gliding motion of the Daleks was inspired by Russian ballerinas. Legend has it that their name came from the spine of an encyclopedia volume labeled "Dal–Lek."

The Dr. Who movies

"The Doctor" is a Timelord who travels through space and time in an English police telephone box, known as the TARDIS. During his adventures, he and his companions take on a variety of hostile beings. Two spinoff movies were made, both starring Peter Cushing, who played the character not as an alien, as he is in the BBC TV series, but as an eccentric grandfather. In *Dr. Who & the Daleks* (1965), The Doctor travels to Skaro and helps the peace-loving Thals to overcome the Daleks, whose reign of terror is conducted from inside a metal city. In *Dr. Who: Daleks Invasion Earth 2150* (1966), the Doctor visits England in the future and foils a Dalek plan to conquer Earth and turn humans into "robomen."

Movie poster, 1965

▲ **Captured on film**
A newspaper cartoon once depicted the Daleks' plans being foiled by a simple staircase, but in *Dr. Who & the Daleks*, they have no difficulty conquering the planet Skaro from their metal city—albeit with the aid of some ramps. In the movies, the Daleks' color and design details vary according to their rank. Their evil leader, Davros, only appeared in the TV series.

▲ **Enemies of The Doctor**
Here, a Dalek is pictured with another of The Doctor's foes. The Cybermen were humanoids who perfected the science of cybernetics. They replaced their flesh with machinery and altered the thought processes of their brains to become emotionless killers.

Ears are based on car turn signals

Head turns to confront the enemy

▲ Hot sellers

Spinoff toys and novelties based on the Daleks remain big sellers, even today. This original 1960s all-plastic toy, made by Palitoy for the BBC, is battery-operated and can also talk.

Eye-stick points from the center of the head

Solar panels surround the body

Gun-stick protrudes from the paneled casing

Blistered body contains a motor drive

Radar rotates with motorized action

Pincers are mechanical and armed with an electrical charge

Arms are retractable

Legs resemble those of Robby the Robot

Base contains belted motors

Robot B9

The Irwin Allen TV series *Lost in Space* featured the heroic robot B9, designed by Robert Kinoshita, the designer of Robby. Like Robby, B9's chest lights flash when he speaks, and his head is full of whirring mechanisms. The youngest member of the Robinson family, Will, is an electronics whiz and is always fixing B9. As a result, the robot becomes the boy's faithful companion, with his famous catch phrases, "Warning, warning!" and "Danger, danger, Will Robinson!"

Lost in Space

The Robinsons are destined to become the first family in space. In the year 1997 (the series began in 1965), they take off in the *Jupiter 2* for Alpha Centauri. Enemy spy Dr. Zachary Smith attempts to destroy the ship, but succeeds only in stranding himself and the Robinsons on a distant world. The series was affectionately known as "The Space Family Robinson," since it was based on the novel *A Swiss Family Robinson*. Known simply as "Robot," B9 lives up to its name ("benign"), and as the series progresses from black-and-white to color, the interplay between him, Dr. Smith, and Will becomes more colorful and comedic.

◄ **Double act**

Two robots were made for the series, costing a total of $75,000. One was the complete motorized figure on tank-style treads, and the other a lightweight, top-only version, which actor Bob May wore. In some shots, the robot appears from the waist up only and moves with an obviously human gait. Dick Tufeld supplied the robot's voice, as he did 30 years later in the 1998 movie remake.

Specification: Robot B9	
Date of TV debut: 1965	
Country of origin: United States	
Studio: 20th Century Fox	
Height: 7 ft (2.16 m)	
Capabilities: Self-navigates, issues verbal warnings, delivers electric shocks	

▶ Spinoffs

This all-plastic, battery-operated 1970s toy is one of a host of B9 spinoffs, and takes considerable liberties with the robot's legs. Many B9 toys and models are poor imitations of the actual robot—here, the image on the box is far more accurate than the toy.

▲ The 1960s cast

Compared to the later movie version, B9 looks positively endearing here, surrounded by his human family. June Lockhart who plays the mother, Mark Goddard, who plays Don West, and Angela Cartwright and Marta Kristen, who play Penny and Judy, have cameo roles in the 1998 remake.

▼ Mean menagerie

Professor Robinson and B9 observe the menagerie of a galactic zookeeper (played by Michael Rennie, the actor who played Klaatu in *The Day the Earth Stood Still*, 1951) who wishes to take Will and his mother away to add to his exhibits.

Hollywood's version

In 1998, *Lost in Space* was remade into a star-studded movie. Reviews were mixed, but the movie made clever use of the original show, playing on the distance between the professor and his son. In the story, a time warp reveals a glimpse of a nightmarish future and forces the father to recognize how little time he has devoted to his son.

Movie poster, 1998

▲ The 1990s cast

The new B9, rebuilt as a heavily armored fighting machine, overwhelms the stars in this publicity shot. William Hurt plays the Professor and Gary Oldman is Dr. Smith. *Friends* star Matt LeBlanc makes his major film debut as Don West.

▼ B9 transformation

The Cyclops-eyed remake of B9 is a sinister figure at the start. But when the robot is destroyed, Will rebuilds it, and the result is a junkyard version of the original show's B9. The new movie uses modern technology to create dramatic special effects.

Huey, Dewey, and Louie

They might be faceless drones on a ship lost in deepest space, but Huey, Dewey, and Louie represent the last vestiges of innocence and goodness as they tend to Earth's remaining plants—and to each other. The quaint trio are the obvious ancestors of *Star Wars'* R2-D2.

▲ **Emotional bond**
The relationship between Freeman and the drones is tenderly depicted, emphasizing their connection, but also their loneliness. In this scene, Huey, Dewey, and Freeman work in the garden.

Silent Running

In this movie, directed by Douglas Turnbull, the spaceship *Valley Forge* carries the remains of Earth's radiation-devastated ecosystem. Ordered to destroy the ship, its captain, Freeman, exhibits the last gasp of humanity's rebellious spirit by refusing to carry out the task. When he and two of his robot companions are lost, it falls to Dewey to be the guardian of the natural world. If *Star Wars* was the defining moment in a decade of film robots spanning the 1970s, the cult classic *Silent Running* surely marks the beginning of the era.

Amazing companions on an incredible adventure...that journeys beyond imagination!

Movie poster, 1971

▲ **Lost companion**
Huey is lost in space as the ship flies by Saturn. The other drones register their loss by gazing into the darkness, giving the movie an emotional depth not normally associated with space sagas.

▶ **Nature lover**
This model of Dewey was made by Vincent C. Backeberg. Dewey was the inspiration behind E.T.'s walk and love of flowers. The drones' innocent quality is emphasized by their names, those of Donald Duck's lovable nephews.

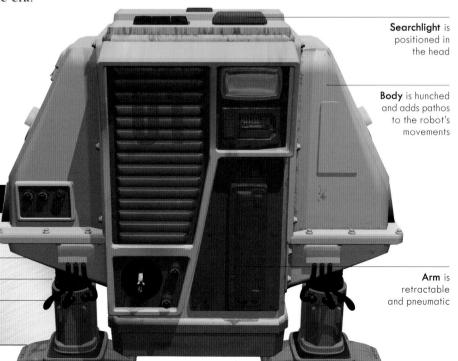

Searchlight is positioned in the head

Body is hunched and adds pathos to the robot's movements

Arm is retractable and pneumatic

Specification: Dewey	
Date of film debut:	1971
Country of origin:	United States
Studio:	Universal Studios
Height:	3 ft (90 cm)
Capabilities:	Walks, plays games, tends plants, protects other drones and human beings

Cylons and Muffit II

The Cylons are the machine despots of a classic TV series that became a film. Robot descendants of a lizardlike race, the Cylons declare war on humanity, whose few survivors flee in a spaceship. Among them is a boy, Boxey, whose playful pet is a cyber version of a Daggit, a cute half-bear, half-dog and the antithesis of the ruthless Cylons.

Battlestar Galactica

The galaxy's human colonies desire a treaty with the Cylons. Instead, the Cylons launch an attack on the 12 planets. The battlestar *Galactica* survives, leaving humanity adrift (an echo of *Silent Running*) in search of the mythical 13th planet, Earth. An unusual band of fighters, renegades, and misfits get together to fight the Cylons as they spread their reign of terror. In 2003, a new TV miniseries recast the Cylons as self-replicating robots who turn on their human masters.

▼ Muffit II, the robot Daggit

This cyber version of a Daggit was created as a replacement for the pet belonging to Boxey, which was killed in the same attack in which Boxey's mother died. Viewers were unaware that Muffit's clumsy movements were those of a chimpanzee in costume.

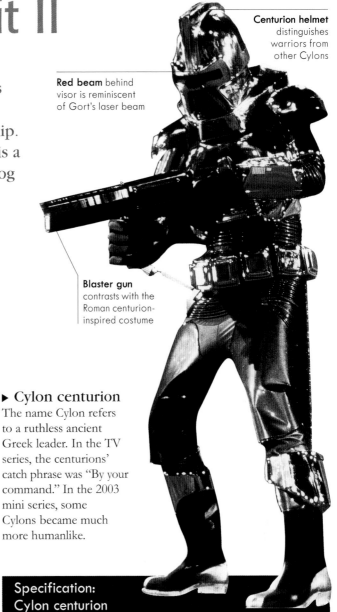

Centurion helmet distinguishes warriors from other Cylons

Red beam behind visor is reminiscent of Gort's laser beam

Blaster gun contrasts with the Roman centurion-inspired costume

▶ Cylon centurion

The name Cylon refers to a ruthless ancient Greek leader. In the TV series, the centurions' catch phrase was "By your command." In the 2003 mini series, some Cylons became much more humanlike.

Eyes are made of yellow glass

Muzzle is made of metal

Joints are metal

Body is covered with fur fabric

Specification: Cylon centurion

Date of TV debut:	1978
Country of origin:	United States
Studio:	ABC-TV
Height:	6 ft (1.8 m)
Capabilities:	Uses two brains, exterminates "unsafe" beings

Specification: Muffit II

Date of TV debut:	1978
Country of origin:	United States
Studio:	ABC-TV
Height:	2 ft 3 in (70 cm)
Capabilities:	Plays and communicates with humans

Star Wars Droids

The original *Star Wars* was an unexpected box-office smash that launched film audiences into a universe of lovable robots and incredible aliens. The real stars of the movie, which went on to become two trilogies, are the droid double act of C-3PO and R2-D2. These robots bicker and rust, but regularly help to save the galaxy, or at least George Lucas's empire. Their debut ignited a new generation's enthusiasm for all things robotic.

Microwave emitter is located on top of the robot's head

The Star Wars films

A long time ago, in a galaxy **far away**, Luke Skywalker (Mark **Hamill**) fulfills his destiny to become **a Jedi knight**, as he, Han Solo (Harrison **Ford**), Chewbacca, C-3PO, and R2-D2 **res**cue Princess Leia (Carrie Fisher) from Darth Vader and the evil Galactic Empire. In the two sequels to the original movie (now renamed *Episode IV: A New Hope*), we learn of the real relationship between Luke, the princess, and Darth Vader as battle is rejoined. The later prequels chart the rise and fall of Jedi prodigy Anakin, Luke's father, as he is lured to the dark side of the Force and his own terrible destiny.

Movie poster, 1977

Shoulder design recalls gladiatorial armor

▶ C-3PO

The delightful, fussy C-3PO comes well-equipped with visual sensing, auditory capability, an olfactory sense, an antenna for receiving other droid messages, and the ability to understand six million galactic languages. In spite of his frenetic ways, he manages to do some good in the end.

Seam on head allows the removal of each panel

Head contains a logic computer

Eyes are photoreceptor units

Mouth is a vocabulator for language interpretation

Button below the chin is an olfactory sensor

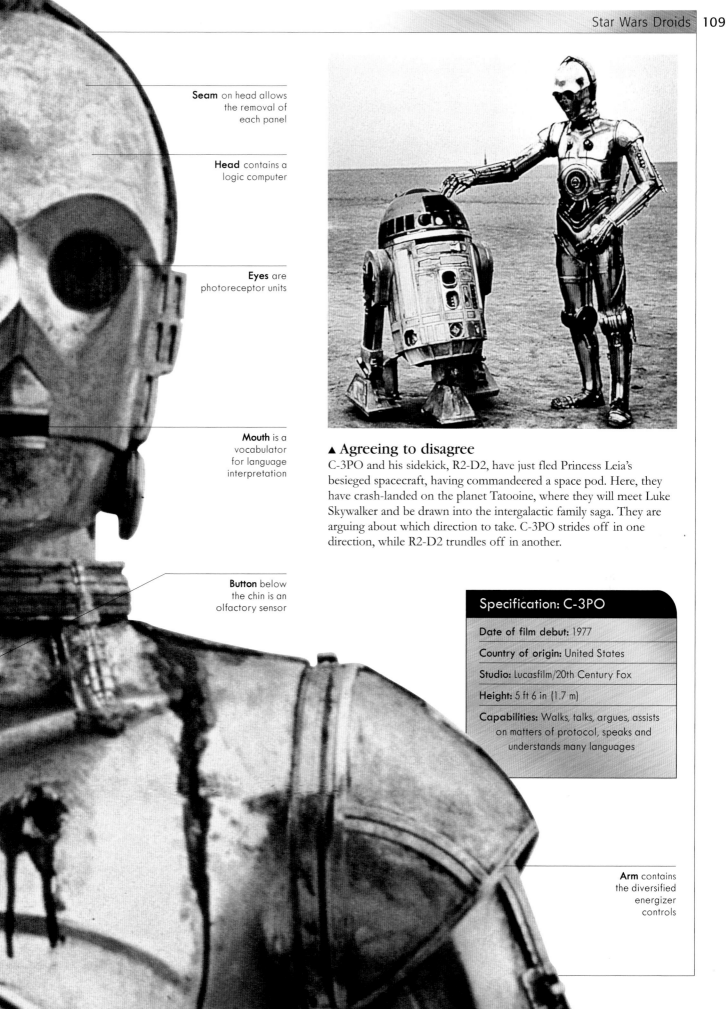

▲ Agreeing to disagree

C-3PO and his sidekick, R2-D2, have just fled Princess Leia's besieged spacecraft, having commandeered a space pod. Here, they have crash-landed on the planet Tatooine, where they will meet Luke Skywalker and be drawn into the intergalactic family saga. They are arguing about which direction to take. C-3PO strides off in one direction, while R2-D2 trundles off in another.

Specification: C-3PO

Date of film debut: 1977

Country of origin: United States

Studio: Lucasfilm/20th Century Fox

Height: 5 ft 6 in (1.7 m)

Capabilities: Walks, talks, argues, assists on matters of protocol, speaks and understands many languages

Arm contains the diversified energizer controls

Perfect partnership

The somewhat strained friendship between C-3PO and R2-D2 created a new benchmark in our perception of robots. They are the classic odd couple, one a humanoid protocol droid and the other a three-legged utility droid. C-3PO excels in languages, while R2-D2 is adept at maintaining and fixing machinery. C-3PO vacillates, while R2-D2 remains calm, even when risking destruction to achieve its mission.

▶ R2-D2

R2-D2's name is short for "reel 2, dialogue 2," a film reel that director George Lucas was once asked for on set. In the original movie, R2-D2 carries the message from Princess Leia to Obi-Wan Kenobi that sets off the chain of events leading to the fall of the Empire. Shown here is a full-size working R2-D2 model, made by robot expert John Rigg.

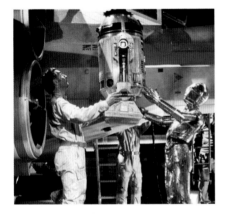

▲ Ready for action

Here, R2-D2 is being prepared to go into battle with Luke and the rebels when they attack the Death Star.

Specification: R2-D2

Date of film debut:	1997
Country of origin:	United States
Studio:	Lucasfilm/20th Century Fox
Height:	4 ft (116.8 cm)
Capabilities:	Walks, carries messages, communicates via beeps and whistles, repairs machines and other robots

Radar system is located in the top of the head

Eye contains the primary photo receptor and radar

Holographic projector is concealed in the head

Chest contains the loudspeaker system

Lower body contains the recharger

Feet contain the motorized all-terrain treads

A cast of thousands

In the original trilogy, there are droids by the dozen to match the menagerie of aliens. Most were played by actors in mechanical costumes. In the trilogy of prequels, there are entire droid armies, mostly created by computer-generated imagery (CGI)—a technology that allows the cloning of thousands of robots for less money than the manufacture of a single one. The *Star Wars* movies also turned spinoff products into successful money-makers. Merchandising includes clothes, masks, toys, kits, figures, games, CDs, websites, comics, books, DVDs, and videos.

◀ Movie posters
While the posters cannot express the breadth and complexity of the stories, they have become icons of graphic design. Some of the original posters are now valuable collectors' items.

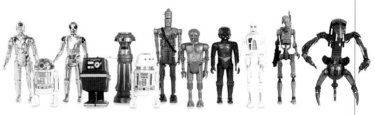

▲ *Star Wars* toys
Almost every one of the legion of droids featured in the movies is available in toy form. These include C-3PO, R2-D2, power droids, assassin droid, battle droids, medical droid 2-1B, and the skeletal 8-D8.

▲ *Star Wars* (1977)
In the original movie, Luke Skywalker rescues R2-D2 and C-3PO from being sold into the robot slave trade on his home planet, Tatooine. The captured R5-D4 watches from the background.

▲ *The Empire Strikes Back* (1980)
Left: The droid K-3PO in the ice cave on planet Hoth.
Center: An Empire surveillance probe seeks out the rebels on Hoth.
Right: The medical droid 2-1B comes to Luke's aid.

▲ *Return of the Jedi* (1983)
R2-D2 and C-3PO approach Jabba the Hutt's hideout on Tatooine, in a shot that echoes the Yellow Brick Road from the *Wizard of Oz*. Like the Tin Man before him, C-3PO embarks on a long and perilous journey.

▲ *The Phantom Menace* (1999)
Battle droids stand in attack formation—these are part of the arsenal of new droid weapons and characters that George Lucas created for the later trilogy using computerized cloning technology.

◀ *Attack of the Clones* (2003)
Battle begins on Geonosis. An army of giant four-legged homing spider droids fights to gain control of the Confederacy of Independent Systems. Each of these droids is equipped with a lethal laser cannon.

Episode III (2005)
The third and final episode in the *Star Wars* prequel trilogy shows Anakin Skywalker dabbling with the Dark Side of the Force. This forces him into conflict with Obi-Wan Kenobi. In the course of the movie, the ultimate destiny of the Old Republic is unraveled.

K-9

A pun on the word "canine," K-9 is the talking robot dog that assists the Doctor, the space-traveling Timelord in the long-running British TV series, *Doctor Who*. K-9 is part faithful pet, part blaster weapon, and part talking computer. Three versions of K-9 appear in the show; however, only two travel with the Timelord. K-9 Mark II sports a plaid collar that rivals his master's famous scarf.

Doctor Who

K-9 Mark II is the Doctor's companion in a series of adventures called *The Key to Time*. The Doctor (Tom Baker) is instructed by the good White Guardian to find six lost crystals, which combine to become the most powerful object in the universe, before they fall into the hands of the evil Black Guardian.

▼ Faithful pet

With his role as a guard dog and ability to detect intruders as well as to fulfill the role of a pet, K-9 is a worthy predecessor to later robotic pets and animal guards, such as AIBO or Banryu.

▲ Inside the TARDIS

The Doctor, fellow Timelord Romana, and K-9 are seen in the interior of the dimension-bending craft, the TARDIS (Time and Relative Dimensions in Space). The TARDIS was meant to assume a form appropriate to the time and place it visited, but budget constraints forced the BBC to stick with the phone box.

▲ Talking toy

K-9 proved popular with viewers, and there were many spinoff products, such as this K-9 model, which comes with a model of the TARDIS.

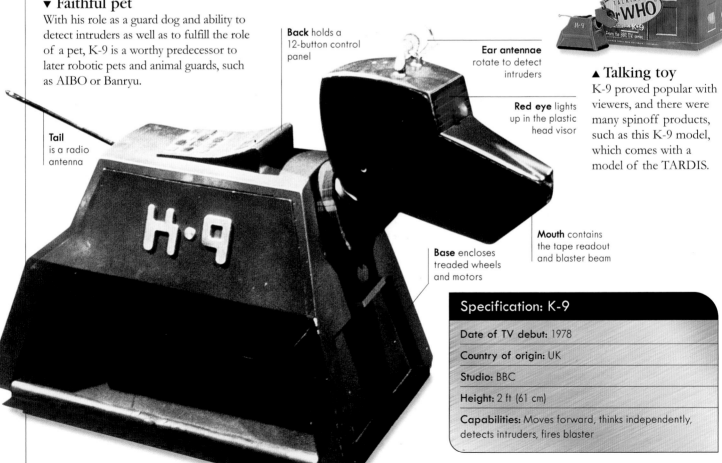

Tail is a radio antenna

Back holds a 12-button control panel

Ear antennae rotate to detect intruders

Red eye lights up in the plastic head visor

Mouth contains the tape readout and blaster beam

Base encloses treaded wheels and motors

Specification: K-9	
Date of TV debut:	1978
Country of origin:	UK
Studio:	BBC
Height:	2 ft (61 cm)
Capabilities:	Moves forward, thinks independently, detects intruders, fires blaster

Metal Mickey

Cult TV favorite Metal Mickey is the malfunctioning metal man built to assist his creator's family, but he is more interested in wreaking havoc. The rotund robot, with a tea kettle for a head and disco lights in his chest, debuted in 1980 on the British children's TV show *The Saturday Banana*. Soon he had his own show, to the delight of children who saw something of themselves in him.

Lovable robotic rogue

The series follows the antics of a robot built by a computer geek to do chores around the home. However, glitches in his circuitry mean that he has a tendency to summon aliens, drink, transport people in time, and play pop music instead of cleaning. The series was directed by *Monkees* star Mickey Dolenz.

▼ Fun-loving handyman

With his catch phrase "boogie boogie"—his response to most situations—Metal Mickey creates mayhem and comedy wherever he goes. As the lyrics to his theme song say: "He's a lot of fun, he weighs half a ton..."

▼ Family pet

Much of the action takes place in the Wilberforce home, where the family becomes the victim of Metal Mickey's antics. Here he is seen with Mrs. Wilberforce, the mother of the family, who was played by Georgina Melville.

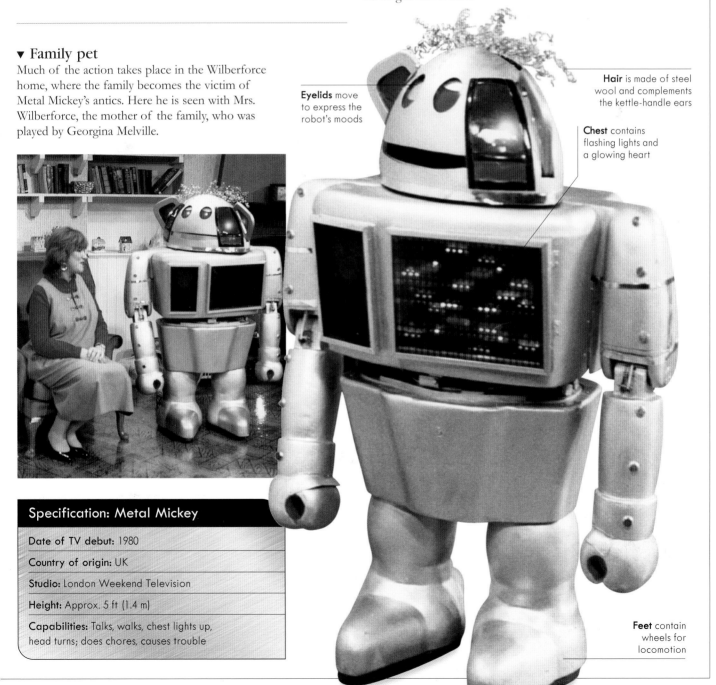

Eyelids move to express the robot's moods

Hair is made of steel wool and complements the kettle-handle ears

Chest contains flashing lights and a glowing heart

Feet contain wheels for locomotion

Specification: Metal Mickey	
Date of TV debut:	1980
Country of origin:	UK
Studio:	London Weekend Television
Height:	Approx. 5 ft (1.4 m)
Capabilities:	Talks, walks, chest lights up, head turns; does chores, causes trouble

Marvin

Marvin is the paranoid android in the TV series of *The Hitchhiker's Guide to the Galaxy* who has a "brain the size of a planet" but remains downbeat. The robot takes every chance to complain, proving that androids with feelings can be as just miserable as some humans.

The Hitchhiker's Guide to the Galaxy

One Thursday Earth is demolished to make way for a hyperspace bypass. Arthur Dent gets embroiled in intergalactic mayhem when his best friend, Ford Prefect, turns out to be an alien. When they are rescued from almost-certain death by the spaceship on which Marvin works, the robot's doom-laden commentary becomes a counterpoint to the characters' struggle to understand the universe. The 1970s BBC radio series by Douglas Adams became a TV hit and bestselling series of books.

◀ **Deep thinker**
Earthman Arthur Dent, played by Simon Janes, finds himself traveling the galaxy in his dressing gown. Marvin claims it gives him a headache just trying to think down to Arthur's level.

Specification: Marvin
Date of TV debut: 1981
Country of origin: UK
Studio: BBC
Height: 6 ft (1.8 m)
Capabilities: Knows everything

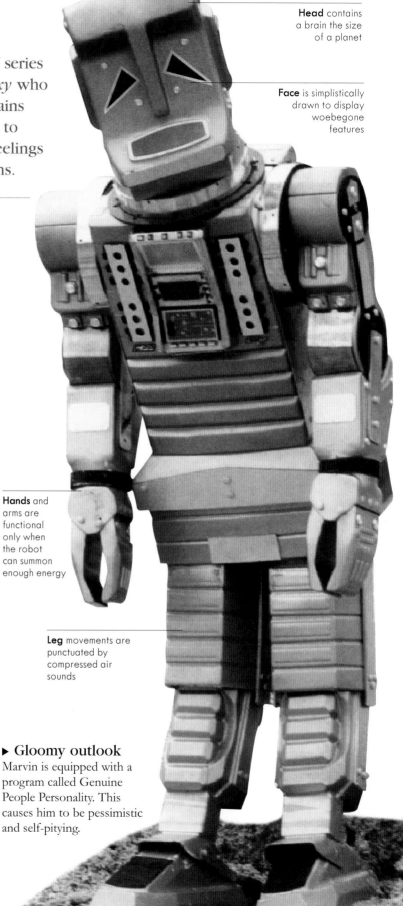

Head contains a brain the size of a planet

Face is simplistically drawn to display woebegone features

Hands and arms are functional only when the robot can summon enough energy

Leg movements are punctuated by compressed air sounds

▶ **Gloomy outlook**
Marvin is equipped with a program called Genuine People Personality. This causes him to be pessimistic and self-pitying.

Kryten

In the UK cult TV comedy *Red Dwarf*, Kryten is the Jeeves-style robot butler, or cleaning and maintenance mechanoid, whose tact and good manners are stretched to their limits by the antics of the less sophisticated spaceship crew. Kryten shot to fame in Series Three, triggering the show's success over five more series.

▲ **Motley crew**
Kryten (Robert Llewellyn) sits beside Cat (Danny John-Jules), holographic projection Arnold Rimmer (Chris Barrie), and Lister (Craig Charles).

Red Dwarf

Curry-loving slob Dave Lister awakes from a two-million-year-nap onboard the spaceship *Red Dwarf* to discover that his only remaining companions are Kryten, the accident-prone robot, a lazy computer, a hologram of the ship's technician, and a highly evolved descendant of the ship's cat.

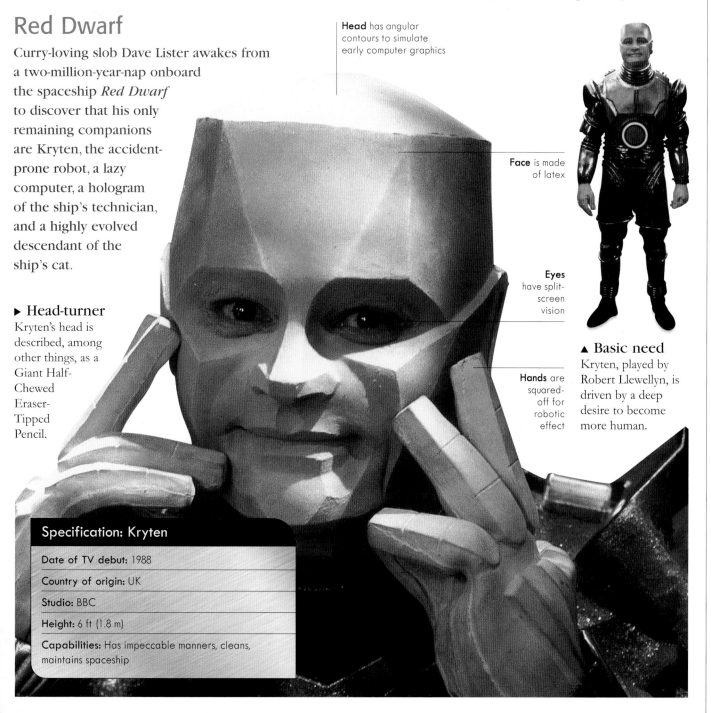

Head has angular contours to simulate early computer graphics

Face is made of latex

Eyes have split-screen vision

Hands are squared-off for robotic effect

▶ **Head-turner**
Kryten's head is described, among other things, as a Giant Half-Chewed Eraser-Tipped Pencil.

▲ **Basic need**
Kryten, played by Robert Llewellyn, is driven by a deep desire to become more human.

Specification: Kryten

Date of TV debut:	1988
Country of origin:	UK
Studio:	BBC
Height:	6 ft (1.8 m)
Capabilities:	Has impeccable manners, cleans, maintains spaceship

Terminator T-800

With *Terminator*, Hollywood created the ultimate robot nightmare: a killing machine with superhuman strength and no human weaknesses. It sees itself as a replacement for humans in a future dominated by artificial intelligence and indestructible engineering. Its mission, if it succeeds, would mean the triumph over humanity of the machines we have created.

The Terminator

The 1984 movie, directed by James Cameron, depicts a future in which human-made machines have become intelligent and self-determining. A humanoid robot, Terminator T-800, is sent back in time by Skynet, the robots' computer network, to change the course of history. His mission is to kill Sarah Connor, the mother of John Connor, the man who will lead humanity's resistance against the machines. The robots must alter the past to ensure that one man never lives, but ultimately they prove to be no match for human heroism.

Forehead seam gives the skull a menacing expression

Eyes can be removed for either repair or replacement

▲ **Unofficial fan club**
There are few official Terminator robot toys, but plenty of figures aimed at the adult collectors' market. These *T3* spinoffs include two T-X endoskeleton figures, and a T-850 in his trademark leather jacket on a stolen police motorcycle.

Teeth are human

Specification: T-800
Date of film debut: 1984
Country of origin: United States
Studio: MGM
Height: 6 ft (1.83 m)
Capabilities: Has superhuman strength and vision, imitates voices, self-repairs, kills without conscience

▲ **The man machine**
This publicity shot shows one half of T-800 covered in engineered living flesh, while the other half shows the robot beneath the skin.

Endoskeleton is made of titanium

◄ **Beneath the skin**
Stripped of its living tissue, the original T-800 is revealed to be a metal skeleton with a grinning death's-head skull. In the third movie, we discover that the human face is that of a marine who volunteered to help the then-human-run company that designed the T-800.

Temple is cut away to expose the inner skull mechanism

Cased wiring replaces the veins and arteries

Generations of Terminator

In the original 1984 movie, Arnold Schwarzenegger plays the Cyberdyne Systems T-800, whose mission to destroy humanity makes him the antithesis of Isaac Asimov's vision of non-threatening machines designed to help humankind. However, in the two sequels, he is reprogrammed to become Connor's heroic defender against a new generation of killer machines.

▶ **Genuine robot**
The T-1 series were the first generation of robots created by the makers of the T-800. T-1 is the only "working" machine among a cast of digital robots.

◄ *The Terminator*
Arnold Schwarzenegger as the malevolent T-800 shows the scars of battle. When the Terminator is damaged, his eye and face begin to reveal the metal robot that is hidden beneath the living human skin.

◄ *Judgment Day*
In the second movie, the reprogrammed T-800 becomes Sarah and John Connor's protector, as he takes on the metal-morphing T-1000, played by Robert Patrick—shown here in his policeman disguise.

◄ *The Rise of the Machines*
The third movie sees Schwarzenegger's T-850 (a modified and upgraded T-800) take on Kristanna Loken as T-X, the ultimate killer machine with an arm that morphs into lethal weapons.

RoboCop

This invincible machine with a human heart represents the ultimate in law enforcement. The inner conflict between gun-toting RoboCop and his repressed human memories makes him a complex character. But there is a light side: RoboCop's design incorporates elements of Gort's head, C-3PO's body, and Maria's legs—a visual joke that echoes *The Terminator*'s blend of high-tech violence and deadpan humor.

▲ **Model cop**
This static figure can be assembled and painted from an all-plastic kit. The detailing on the model is high-quality and the helmet is detachable.

Futuristic black comedy

This violent and darkly comedic film directed by Paul Verhoeven is set in the crime-ridden Detroit of the future, and opens with police officer Alex Murphy (Peter Weller) mortally wounded in a shootout with villains. Murphy is rebuilt as a cyborg by the sinister OmniConsumer Products Corporation (OCP), a corrupt organization that controls local law enforcement. While RoboCop is hailed as the cutting edge of modern policing, Murphy himself is tormented by fragmented memories of his human past. As he attempts to piece together the events of his previous life, he discovers that his cyber-enhanced fight against crime and his quest to track down those responsible for his new situation may actually be one and the same thing.

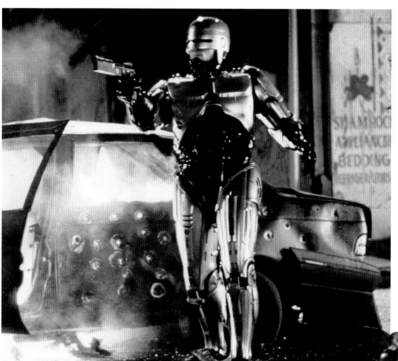

◀ **Formidable foe**
"Part man, part machine, all cop" was the tagline for the original movie, revealing RoboCop to be an unstoppable crime-busting tour de force.

Movie poster, 1987

▶ Recharging

RoboCop sleeps in a chair that recharges his batteries and reconfigures his programming. But when he is in sleep mode, Murphy discovers that he is haunted by memories of his human past.

▼ Shedding weight

RoboCop reveals what is left of the human beneath the robotic visor. In the first movie, the suit was made of fiberglass and took several hours to put on. In later films, it was much lighter and could be put on more quickly.

Ear connects to audio record and playback system

Face is the only remnant of the human form

Armor is bulletproof

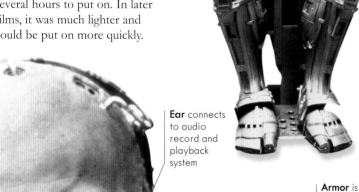

Specification: RoboCop

Date of film debut: 1987	
Country of origin: United States	
Studio: Orion Pictures Corporation	
Height: approximately 6 ft (1.8 m)	
Capabilities: Fights crime, deflects bullets, records and plays back sounds, shoots with deadly accuracy	

The RoboCop Trilogy

The original movie was a box office hit. The two sequels are more violent, but they lose some of the sharp scripting and satirical edge. In the second movie, RoboCop defends Detroit against a new drug and a criminal robot, while the final installment sees him fighting for people displaced by the city's corrupt rulers.

▲ RoboCop

The huge ED-209 robot is introduced by OCP as the ultimate in crime prevention, but its lack of human subtlety makes it no match for RoboCop.

▲ RoboCop 2

This time Murphy faces a drug dealer killed in a raid, who has been rebuilt into RoboCop 2, a superior cyborg. It also retains human memories, but only criminal ones.

▲ RoboCop 3

Murphy confides in fellow officer Anne Lewis (Nancy Allen), who has befriended him. Weller is replaced by Robert Burke in this film, directed by Fred Dekker.

Dangerous droids

SCIENCE FICTION often portrays robots as personifications of evil, as destructive machines that possess intelligence and strength but not human conscience. In his books, Isaac Asimov's Three Laws of Robotics were intended to rein in the machines, but cinema's countless rampaging robots and cyber psychos demonstrate the possible repercussions of the machines reigning over us.

▲ **The Devil Girl from Mars**
The towering alien robot Chani looks like a refrigerator but can incinerate its human foes, in this 1954 movie set on the Scottish moors.

◀ **King Kong Escapes**
Heroic King Kong battles the evil robot Mecha Kong and a host of legendary monsters in the 1967 Japanese movie.

Since storytelling began, tales of man's futile attempts to rival the power of God have always showed how he was ultimately destroyed by his arrogance. Whether it is a person sacrificing humanity for superhuman knowledge, as in the *Faust* legend, or bringing life to dead matter, as in *Frankenstein*, human beings are always punished for their folly when the results of their actions return to haunt them. Robots are human-made creations, and many robot

▶ **Colossus of New York**
The 1958 movie features a killer robot with a scientist's brain.

Stature suggests superhuman physical power

Rusting metal betrays the robot's age

▼ **Judge Dredd**
The 1995 movie features a ruthless human law enforcer battling an antique robot that has been recycled to kill again.

folly. A *Frankenstein*-inspired robot with a scientist's brain leaves a trail of devastation in *Colossus of New York*. In *Westworld*, scientists find themselves unable to control their robotic theme park attractions. In the 1979 movie *The Black Hole*, a killer robot, Maximillian, turns on its mad creator as it patrols a ship that is falling into the symbolic chaos and destruction of the void.

A mission to destroy

Human beings can only cower in terror when droids go on the rampage. In *Saturn 3*, the evil, headless robot, Hector, represents a terrible, faceless power as he sets out to pillage, and to destroy a happy family. Even the forces of law and order cannot protect us from the robots' onslaught. In *Judge Dredd*, a rusting, recycled robot is still more powerful

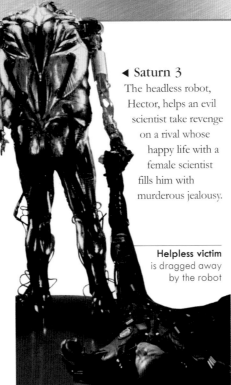

◄ **Saturn 3**
The headless robot, Hector, helps an evil scientist take revenge on a rival whose happy life with a female scientist fills him with murderous jealousy.

Helpless victim
is dragged away by the robot

▼ **The Black Hole**
A mad scientist creates a lethal armored robot called Maximillian in the 1979 Disney film *The Black Hole*.

recycled robot is still more powerful than the law. The relentless robot in *Terminator* leaves a trail of burned-out police cars and death behind him in his mission to kill one person. In the sequel, *Terminator 2: Judgment Day*, a shape-shifting robot strips away every line of human defense by assuming the form of a policeman and imitating the voices of innocent people he has killed.

Man versus the machines

While the good robots strive to be human or to exhibit human characteristics, their evil counterparts remain machines through and through. Devoid of human "flaws," such as mortality, remorse, or pity, evil robots can wreak havoc without considering the consequences—and this is what makes them so terrifying. The forces of good might outsmart them in the final reel—even when the odds are stacked against humanity—but box-office receipts rely on its being a fight to the bitter end.

Human disguise

In movies, robots seem to be at their most dangerous when they closely resemble their creators. By assuming human form, these evil doppelgangers move unnoticed among humanity, only to turn on their creators in the ultimate battle for survival: humankind versus our machine replacements.

Ash is about to be revealed as a robot as he battles the heroic Ripley

▲ **Robot revealed**
In *Alien* (1979), science officer Ash admires the Alien's lack of human conscience. His attitude is explained when he is revealed to be a robot who is prepared to sacrifice the crew to ensure that the Alien survives.

▲ **Death's messenger**
In the 1973 movie *Westworld*, theme park visitors believe the Cowboy robot (Yul Brynner) is acting for their amusement, only to find he is a genuine killer.

Data

As the hyperintelligent robot that wants to be human in *Star Trek: The Next Generation*, Data has many adventures revolving around his attempts to understand human emotions, while offering the skills of a machine. Data's powerful brain is described as "positronic," a term from Isaac Asimov's *I, Robot* stories. He is equipped with an emotion chip, which makes him self-aware when it is activated.

Star Trek: The Next Generation

This series, which began in 1987, echoes the original *Star Trek* (first broadcast in 1967) only in a few nostalgic details—the name of the ship *Enterprise*, the order of command, and the dress code. Set in the twenty-fourth century, Captain Jean Luc Picard (Patrick Stewart) and his crew, including Data (Brent Spiner), travel through space in a far larger and more up-to-date *Enterprise*. The use of computer animation, unavailable to the original series, allows the crew to encounter more believable new planets and cultures, including a number of robotic friends and foes.

▲ **The new-generation crew**
The new *Enterprise* shares some characteristics with the one in the original series—for example, the crew still uses the famous transporter room to beam down to planets.

▶ **Sherlock Data**
The new starship has a "holodeck" where the crew can take part in virtual reality adventures. In one episode, Data fulfills an ambition to "become" Sherlock Holmes and battle a holographic Moriarty.

Star Trek's malevolent machines

Evil robots have been a staple of movies and TV since Maria of *Metropolis*. Data is compromised when his ethical programming is turned off by his twin, Lore, in the episode *Descent, Part II*. Lore is Data without the redeeming influence of human society, and comes from a long-established science-fiction tradition of evil twins. This plot device can be traced back to the original series when Captain Kirk was duplicated in the episode *The Enemy Within*. The Borgs in *The Next Generation* are good examples of the perils of cybernetics. Part of *Star Trek's* legion of evil robots, they conduct a technology-enhanced quest to absorb all cultures into their own, and consider individuality an imperfection.

▲ **Destructive Borgs**
The Borgs live to assimilate and destroy, using technology to absorb the wisdom of entire races into their consciousness.

▲ **Lore, the evil twin**
The character of Data's evil twin, Lore, is in many ways opposite. Where Data is rational and in control, Lore is emotional and sadistic.

Head houses a brain more powerful than the *Enterprise* computer

Eyeballs are yellow to show that he is not human

Skin has a golden tone

Uniform is that of a lieutenant commander

Specification: Data

Date of TV debut: 1987

Country of origin: United States

Studio: Paramount Studios

Height: 5 ft 11 in (1.7 m)

Capabilities: Experiences human emotions, has super-powerful brain and superhuman vision

◄ Hero android

In the storyline of *The Next Generation*, Data was created by Dr. Noonien Soong in the year 2335. He is a lieutenant commander aboard the *Enterprise* and is known to have a brother (Lore) and sister, as well as a pet cat named Spot. Data has been highly decorated for his services, and his honors include the Starfleet Command Decoration for Valor and the Medal of Honor.

Bender

Star of the TV cartoon series *Futurama*, Bender is a wisecracking, foul-mouthed antihero whose contempt for humans knows no bounds. Designed to bend girders, he devotes his time to bending the rules and testing the limits of interplanetary politeness.

Specification: Bender	
Date of TV debut: 1999	
Country of origin: United States	
Studio: 20th Century Fox	
Height: 5 ft 9 in (1.8 m)	
Capabilities: Bends girders, cooks, fights, smokes, drinks, swears, cheats	

Futurama

The cult TV show is the brainchild of *The Simpsons'* creators, Matt Groening and David X. Cohen. Bender the robot is rescued from Robot Hell by Fry, a pizza-delivery boy from the twentieth century who finds he has woken up in the thirty-first century. Bender has a taste for Olde Fortran Malt Liquor (Fortran is the name of an early computer-programming language) and for Mom's Old Fashioned Robot Oil.

Losing his head

◀ ▼ **Colorful character**
Bender, seen left in a parody of the classic astronaut pose, teams up with Fry and one-eyed alien girl Turanga Leela (below, on the scooter). Together with a cast of bizarre characters, they have many adventures. Bender gambles, goes computer dating (by dating a computer), and confronts his evil nemesis (who looks exactly the same as Bender, but with a beard).

Bender fights for the love of a fembot

Bender and the Harlem Globetrotters

Bender as a "werecar" at full moon

Fry and Leela whiz around the rooftops of 31st-century New York, passing Planet Express, where Bender works as a ship's cook

▶ **Factory-produced**

Bender was born in a factory in Mexico, and he needs chemical energy from alcohol to function properly. Tragedy struck his family when his father was killed by a can-opener.

Antenna is for interplanetary communication

Eyes have pixel-style square pupils

Teeth are arranged in three layers to aid cigar-chomping and cursing

Door conceals beer and stolen goods

Arms are designed to bend girders

Legs are unsteady due to excess alcohol intake

THE GENDER BENDER

▲ **Bender toys**

Among the merchandize available are these two clockwork toys. One chomps a cigar and holds a beer, while the other wrestles with an identity crisis in a blond wig and a dress.

▲ **Escape from Hell**

In this early episode of the show, Bender finds himself trapped in Robot Hell—the dreadful place where unwanted or misbehaving robots go—before he is eventually rescued by Fry.

Little Robots

An animated TV show for preschool children, *Little Robots* is based on the book by Mike Brownlow and has a cast of lovable robots with an environmentally friendly message. Each robot has a different personality with its own strengths and weaknesses, and the key message is that "every little robot is good at something."

▲ **Faithful friends**
Each of the robots has a very different personality, such as the thoughtful Tiny and the mischievous dog Messy, but they learn to forge firm friendships.

Building a happy home

The story centers around a group of robots that find themselves abandoned in the middle of a junkyard full of nuts and bolts and other discarded metal objects. Using skill, vision, imagination, and a pioneering spirit, they create their own home, complete with a sun, moon, and trees. The pertinent lessons in cooperation and ecological sense are delivered with a dose of humor.

Hands wear baseball gloves

Shoulders have gladiator-style armor

Funnel lets out steam when Rusty panics

Wings are batlike to reinforce scary persona

▶ **Gang of friends**
The cast of characters all contribute something different to the team of robots. They include the hyperactive Sporty, the little good guy Tiny, the magician Scary, Noisy, who never stops performing, Rusty, who is very accident-prone, and Messy the dog, who hates being clean.

Belt houses Tiny's controls

▶ Little LEGO robots

LEGO has created a line of toys based on the TV characters. These plastic toys stick faithfully to the look of each character and are nearly as pliable as the animated versions.

Chest buttons play "Sporty" sounds

Head contains Tiny's tool kit

Skateboard clips to the toy's feet

▲ Community spirit

The little robots discover that by working together as a team and sharing their skills, they can build a home for themselves from junk.

Antenna resembles a tennis racket

Specification: Sporty

Date of TV debut: 2002

Country of origin: UK

Studio: Create TV & Film Ltd.

Height: 10¾ in (27 cm)

Capabilities: Works out, plays tennis against himself, motivates the team

Antenna is large to receive lots of clever ideas

▶ Opposites attract

Tiny and Sporty are a duo who play off against each other. Tiny is creative and brainy while Sporty is all muscle and action. Each would like the other to become enthusiastic about their individual passions. Their friendship reflects the fact that contrasting personalities are often drawn to each other in real life.

Chest plate has a control panel

Knees are protected with sports-style kneepads

Feet have a wing motif

Specification: Tiny

Date of TV debut: 2002

Country of origin: UK

Studio: Create TV & Film Ltd.

Height: 6½ in (16 cm)

Capabilities: Solves problems, fixes things, operates day/night lever

Children's favorites

ROBOTS PLAY A LEADING role in many children's movies and TV series. Some are heroic defenders of children against the forces of evil, or sympathetic outsiders in need of a friend; others are mischievous characters with whom children identify. Paradoxically, robots often offer the love and guidance that adults have failed to provide.

Computer
hangs around the robot's neck

Many robot stories are fairy tales in all but name, with robots displaying values such as loyalty and heroism that the flawed human characters frequently lack. One of the great modern-day robot fables is the poet Ted Hughes' *The Iron Man*, which was adapted as the feature-length cartoon, *The Iron Giant*, in 1999. In this classic tale, a rusting robot comes to represent the old-fashioned virtues of nobility and

self-sacrifice, much like the medieval knight he resembles. The story takes place in a Cold War environment, casting adults as remote, paranoid characters in the mind of a boy the robot befriends.

Robotic magic

Fantasy robots also have something of the wizard about them—they can summon up "magical" powers. Among these are superhuman strength and

▲ **Buck Rogers in the 25th Century**
Twiki is the catch-phrase-mumbling robot that subverts Buck Rogers' grown-up world of guns, battles, spaceships, and relationships.

▶ **Batteries Not Included**
Robotic flying saucers come to the aid of a couple in this 1987 movie; forerunners of today's intelligent appliances, perhaps.

speed, or the ability to transform (a myth descended from fairy tales). Robots like these are seen by children as faithful protectors, often taking the place of adults whose busy lives take them away from home in the story. In the computer-animated TV series *Cubix*, an old, discarded robot is

Cubix will go to any length to protect his friends

▼ **Cubix**
In the cartoon series, Cubix, the children, and other robots live in Bubble Town, where the top robot manufacturer, RobixCorp, is based.

◀▼ **The Iron Giant**
This 1999 movie features a giant robot who befriends a young American boy at the height of the Cold War.

IT CAME FROM OUTER SPACE!
THE IRON GIANT
AUG 6

restored by a boy. It turns out to be a transforming robot guardian who runs, flies, or rolls to the rescue, never letting him down—even when adults do.

Humanizing the machine

Computer games allow children to experience the illusion of super-powers as they fight virtual battles with all-conquering robot allies. In this way, robots have increasingly become seen as powerful friends in children's imaginations—friends whose size or appearance might be strange, or even frightening, but who are accepted and respected for who they really are. This trend is reflected in many movies, cartoons, and TV series where robots play a leading role. In such stories, children usually give the robots something in return: human emotions, friendship, and love.

Tin men with a heart

The definitive example of children helping to humanize robots can be found in the *Oz* stories by L. Frank Baum, in which a clockwork man and a tin man are as noble as any hero, and through friendship with Dorothy, the latter learns he has a heart. Similarly, in the 1986 movie *Short Circuit*, a robot defense drone is hit by a bolt of lightning—a symbolic act of creation. This transforms him from a machine into a being with a conscience who is then accepted into the family home. Indeed, this identification between robots and children has become so strong that the robots are often the characters with whom children identify the most in a movie, especially if the

▶ **Metal pal**
Although not referred to as a robot, Tik Tok is Dorothy's clockwork friend in the 1985 movie based on the *Oz* books.

Body resembles recycled industrial machinery

rest of the human cast are portrayed as boring adults once they have set their ray guns aside. One such robot is Twiki, the sidekick of the human hero in the 1970s TV series *Buck Rogers in the 25th Century*. He is a squeaky-voiced, child-sized robot with a hint of anarchy about him.

Twenty-first century toys

However, the world is changing, and many children now have real robot toys to play with. For them, robots are no longer a myth, a cartoon, a special effect, or a fairy tale—they are a reality.

The Wizard of Oz

The character Dorothy in L. Frank Baum's much-loved *Wizard of Oz* books was a heroic child in a strange and threatening world. Tik Tok and the Tin Man are robots in all but name, and their friendship with Dorothy is the blueprint for many subsequent robot tales.

◀ **Collectors' item**
This replica of the original Tin Man, complete with his hard-won heart, is aimed at the collectors' market.

Features are simplified to appeal to children

▶ **Celluloid tin man**
In the 1985 movie *Return to Oz*, the Tin Man is played by an articulated puppet.

No. 5 waves to his new-found human friends

Cowboy outfit suggests an old-fashioned hero

▶ **Short Circuit**
This 1986 movie stars a wheeled defense robot, SAINT No. 5, which comes to life when it is struck by lightning.

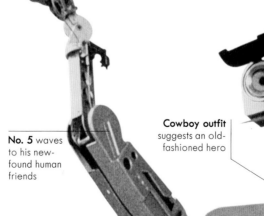

Computer Games

A legacy of robots in comics, movies, TV, and toys inspires many of the computer games that are popular today. Players identify with computer-generated characters populating cyber-worlds with breathtaking graphics and sound effects.

TransFormers

In a battle for control of the planet Cybertron, players in this game can "be" one of the heroic Autobots fighting the evil Decepticon robots. Both sides are seeking the aid of the Mini-cons, a race of tiny Transformers. Like their toy counterparts, the Autobots can change into vehicles.

PlayStation packaging

▶ Robot evolution

To create a game character such as Hot Shot, a wireframe model is computer-generated, then covered in a "skin." Finer details further define the character, while the story brings its personality to life.

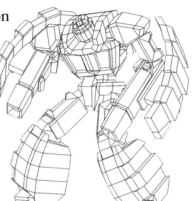

Optimus Prime strides forth

Hot Shot destroys an enemy Decepticon

Specification: TransFormers	
Date: 2004	
Publisher: Atari	
Developer: Melbourne House	
Platform: PS2	

◀ Larger than life

Optimus Prime and Hot Shot are on Earth to fight the evil Decepticon army. Optimus (far left) transforms into a truck, while Hot Shot becomes a sports car.

Optimus Prime leaps into action

Hot Shot takes aim

Robotech Battlecry

The gigantic aliens the Zentraedi are invading Earth to subjugate the human race, but they have a formidable adversary in Jack Archer, pilot of the versatile Veritech. This machine can transform into a giant mechanized robot, a high-flying jet fighter, or a hybrid unit that has elements of both. Players must use more than one mode to complete their objective, and the game has 45 missions— enough to keep the battle going forever.

XBOX packaging

Specification: Robotech	
Date: 2002	
Publisher: TDK Interactive	
Developer: Vicious Circle	
Platform: PS2, Gamecube, Xbox	

▼ **Changing backdrops**

The stunning graphics of Robotech Battlecry provide the player with richly detailed backgrounds to the action. The battles are set in city ruins, canyons, the skies above Macross City, and space.

Giant Battleoid wields a missile launcher

Veritech fighter battles high above Earth

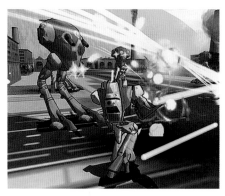

Fighter confronts the Zentraedi

Armored Core 3

In this saga, a devastating war has wiped out most of human life on Earth. The Ravens are mercenary pilots who fight for the highest bidder. Currently, they are fighting on behalf of Earth's corporations, while most of what remains of humanity cowers in a large city underground. In the hundreds of different scenarios, the player controls some of the huge robots, the Mechs, that both sides in the battle are using to destroy each other.

PlayStation packaging

Specification: Armored Core 3	
Date: 2002	
Publisher: Metro3D	
Developer: From Software	
Platform: PS2, Gamecube, Xbox	

▼ **Dramatic scenes**

Most of the action takes place in a cityscape. The camera angle can tilt up and down, and also zoom in and fade out on the background for dramatic effect.

A Raven comes under attack

A Raven rampages through the city streets

A Raven with its arsenal of weapons

Pulp Illustration

At a time when most households did not own any domestic appliances, artists were depicting incredible machines that wielded power over humans on Earth. The age of robot pulp illustration began with the man who coined the term "science fiction," Hugo Gernsback, and his magazines *Amazing Stories* and *Wonder Stories*. He inspired many artists to create images of fantastic robots set in terrifying future worlds.

Influential art

The pulp illustrations of artists such as Frank R. Paul, Ed Emshwiller, and Alejandro Canedo brought science fiction to the masses. This style of magazine art defined the universe of robots, spaceships, and aliens that influenced the work of Arthur C. Clarke and Isaac Asimov and, later, Hollywood movies.

◄ ▶ **Startling Stories**
The January 1939 issue of *Startling Stories* featured Stanley G. Weinbaum's tale *The Black Flame*, illustrated with a diabolical robot keeping human specimens under glass. The towering robot with destructive laser vision wields total control over the tiny, powerless beings.

A NOVEL OF THE FUTURE COMPLETE IN THIS ISSUE!

STARTLING STORIES *JAN.*

15¢

A THRILLING PUBLICATION

FEATURING:
THE BLACK FLAME
A Book-Length Novel of the "Land of Time to Come" By STANLEY G. WEINBAUM
ALSO MANY OTHER STORIES & FEATURES

ALBERT EINSTEIN — HIS LIFE STORY TOLD IN PICTURES!

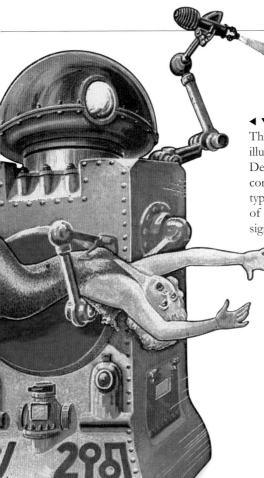

◀ ▼ Wonder Stories

This dramatic kidnap by a robot, as illustrated by Frank R. Paul for the December 1931 issue of *Wonder Stories*, contains an element of humor that is typical of pulp illustration. The impact of Paul's colorful cover art made a significant contribution to the success of Gernsback's magazines.

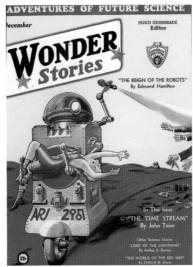

▲ Galaxy

Ed Emshwiller's September 1954 cover for H. L. Gold's *Galaxy* shows a scientist repairing a bionic woman. Galaxy novels were the forerunners of modern paperbacks. A leading science-fiction artist of the 20th century, Emshwiller also had an impact on experimental filmmaking.

▲ Astounding Science Fiction

This 1949 illustration by Alejandro Canedo from John W. Campbell's *Astounding Science Fiction*, titled "Missing Ingredient," shows a robot without a heart, an affliction first portrayed by *The Wizard of Oz's* Tin Man, and later echoed in Isaac Asimov's *The Bicentennial Man*.

▲ Scoops

The first British pulp science fiction magazine, issued in 1934, lasted only a year. The cover of the first issue has echoes of World War I. The second depicts a flying robot while the third has a doomsday robot, illustrating a story by the Dutch writer Desiderius Papp.

Robot Artists

From the Futurists of the early twentieth century to the artists of today, robots and machines have long been a source of fascination and inspiration. Creating robots from the imagination means that there are no technical constraints, so representations of robots in paintings and graphic art can conjure a wide range of images, from dramatic and complex to witty or weird.

Michael Whelan

In his 30-year career, Michael Whelan has illustrated the work of science-fiction masters such as Asimov, Simak, and Saberhagen. Through his work, he can conjure up the good and the bad, the heroic and the hostile. He is a renowned visionary artist whose work can be seen on magazine, book, and album covers around the world.

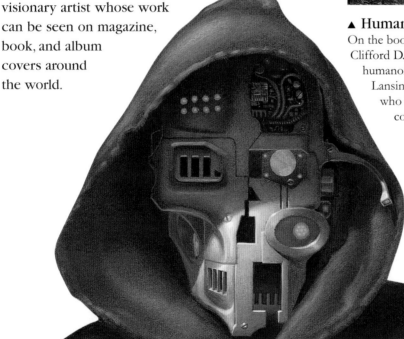

▲ Humanoid on a quest
On the book cover of the 1982 novel *Special Deliverance* by Clifford D. Simak, Whelan depicts a post-*Star Wars*, tea-serving humanoid that keeps company with the protagonist Professor Lansing. The two are part of a group of assorted travelers who have found themselves on a strange planet and must complete a life-or-death challenge.

◀ Hooded robot
Brother Assassin comes from the book of the same name in the popular series of *Berserker* novels by Fred Saberhagen. The evil robot, with his hooded, skull-shaped head and mechanical talons clutching a clock, resembles a macabre Father Time.

▶ Space wars
This painting is a computer-generated image of a battle scene in a bleak future world and is an expression of the theme of war. The stark robot silhouette set against a blazing orange sky helps to create this strong, and rather disturbing, image.

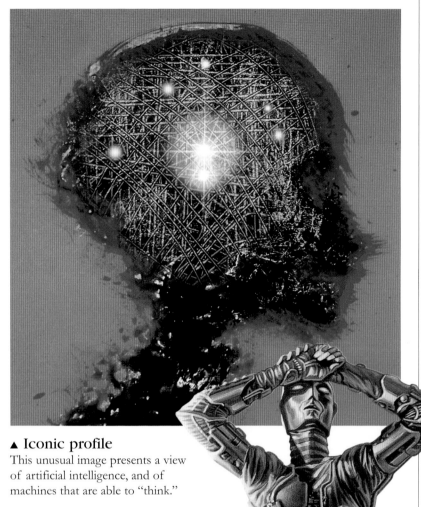

▲ Heroic and hostile

This illustration was used on M.A.R. Barker's novel *The Man of Gold.* The metal robot figure, or Man of Gold, assumes a threatening posture, suggesting that machines might rise to a level of intelligence and power at which they could pose a threat to organic life forms. The use of vivid, unrealistic colors and the inclusion of a dramatic visual story heighten the sense of fantasy and adventure.

▲ Iconic profile

This unusual image presents a view of artificial intelligence, and of machines that are able to "think."

▶ Languid pose

R. (Robot) Daneel, an Asimov character in *The Robots of Dawn,* strikes a very human pose as morning breaks on the planet Aurora.

Eric Joyner

The oil paintings of Eric Joyner echo the heroic figures set in urban landscapes that were painted by US artists from the Ashcan and Brandywine Schools at the turn of the nineteenth century. San Francisco-based Joyner also creates illustrations for books, magazines, computer games, and advertising. He sees his subjects in challenging but romantic settings—the last of the tin toys lost in space.

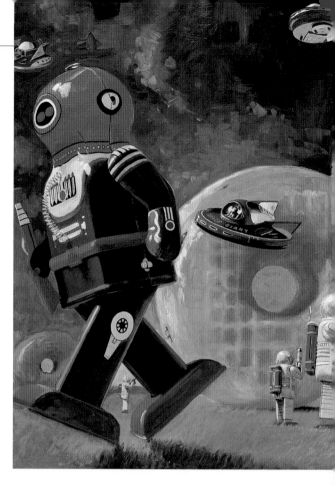

◄ **Another Green World**
Atomic Robot contemplates the universe as he relaxes on a distant planet very much like our own.

▶ **Glazed**
Space Captain Highwheel and others defend Earth from an invasion by giant doughnuts.

▼ **The Final Blow**
Rock 'em Sock 'em robot toys slug it out in the ring as a host of other robot toys look on.

◀ Sparky
Marooned in space, Highwheel Robot hurtles past planets and asteroids. Catastrophes may occur, but progress continues.

▼ Robo Kong
In a pastiche of the famous movie scene, Star Strider Robot sits high above the city, holding aloft a toy plane with a doll pilot as the upstaged gorilla approaches.

◀ The Conqueror
Moon Explorer plants its flag on a barren asteroid, in a comment on how nations race to claim new territories, regardless of their potential uselessness to humankind.

▶ The Last Tin Man
A gigantic Atomic Man stands forlornly in the shadow of a medieval ruin as tourists drive by.

▼ Sunday Cruise
In this detail, Mr. Zerox strikes a laid-back pose as he takes a leisurely drive in the countryside—robots need a day off, too.

Karl Egenberger

These brightly-colored pictures hark back to 1950s optimism about the future world of robots and space exploration. Baltimore-based graphic artist Egenberger depicts classic robot toys within computer-generated spatial landscapes in vivid colors. Using a computer allows Egenberger to achieve a high level of detail and complexity. Most of the models come from his very large personal collection of tin and space toys.

▼ Chief RobotMan

Tin-plated Chief RobotMan stares out of a green mountainous landscape. The original battery-operated toy was produced by the Yoshiya Corporation in the mid-1950s. The skirted body is reminiscent of Matsudaya's Gang of Five.

◀ Smoking SpaceMan

A smoking robot stands against a glowing purple galaxy dotted with candy-colored planets. The original early battery-operated toy was made by the Yonezawa-Linemar Corporation of Japan.

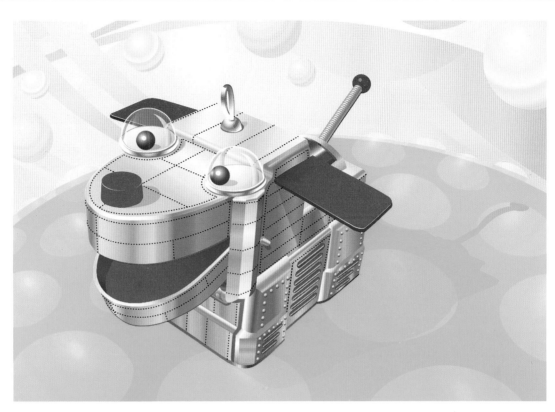

▶ Magic Space Dog

An early 1950's Yoshiya Corporation wind-up toy robot dog flies through colorful bubbles. The painting gives a playful view of space travel before it became a reality.

◀ Directional Robot

With its red eyes glowing, this skirted bump-and-go robot is surrounded by ethereal pearly planets. The original 1950s toy was produced by the Yonezawa Corporation of Japan.

▼ Flashy Jim

An innocent-looking robot toy wanders across a barren, green planet against a starry sky. Flashy Jim was a 1950's remote-control toy made by the Sankei Corporation of Japan.

Robot Sculptors

Technological innovation and the Bauhaus and Kinetic Art Movements of the early twentieth century have inspired robot sculptors such as Clayton Bailey, Christian Ristow, and Lawrence Northey. Their robots express their views on the relationship between people and machines, and range from whimsical to brutal, from small-scale pieces to giant machines the size of bulldozers.

Clayton Bailey

Local flea markets and scrap-metal yards may seem unlikely locations for an artist to source materials, but that is exactly where American sculptor Clayton Bailey finds the discarded appliances, cookware, and car parts that he uses in his work. Since 1976, he has been transforming other people's junk into charming life-size robot sculptures, each with its own quirky character.

Eyes light up inside the head

Nose is shaped like a faucet

Head is made from a metal coffee pot

Hands are sculpted in great detail

Hand is made from a corkscrew

Antenna is a coiled wire

Mouth is set in a grimace

◄ Crackling Robot

This robot is classic Clayton Bailey. It is a life-size standing robot made from crafted aluminum, with beady eyes, and a glass panel in its torso revealing a neon tube that lights up and crackles. This is a robot sculpture with attitude.

Neon tube in chest lights up

◄ Alien Robot 2

Alien Robot 2 has sad, illuminated eyes, a charming bent nose, and a mouth that looks like a harmonica. He is made from aluminum cookware and other found objects, and is held together by rivets—reminiscent of the design of some early tin toys.

Specification: Crackling Robot

Date of manufacture: 2000

Country of origin: United States

Height: 6 ft (1.8 m)

Power source: AC electricity

Features: Neon tube in chest lights up and crackles, eyes light up

Specification: Alien Robot 2

Date of manufacture: 1996

Country of origin: United States

Height: 4 ft (1.2 m)

Power source: AC electricity

Features: Eyes flash when activated by sound

Breasts are made from tin bowls

Legs are made from bent rods

Slippers resemble flatirons

Cup is made from spun aluminum

Head is conical in shape

Visor in helmet lights up

Body is wedge-shaped

Legs are made from thin tubes of steel

Feet are fastened to legs with skewers

▲ Tipsy

Bailey's palette of wit and whimsy is in full use in Tipsy, whose name says it all. This robot figure is made of recycled aluminum, cookware, and various found objects, including a coffee pot, that provides a rather distinguished nose and a haughty air.

▶ Alien Robot

It is unusual for a Bailey robot to be faceless, but this "I await your instructions" posture is typical of much of the sculptor's work. The figure, bolt upright on its very long legs, has an immediate relationship with its environment, and commands attention.

Specification: Tipsy

Date of manufacture: 1996

Country of origin: United States

Height: 3 ft 9 in (1.1 m)

Power source: None

Features: Freestanding static sculpture

Specification: Alien Robot

Date of manufacture: 2002

Country of origin: United States

Height: 3 ft 9 in (1.1 m)

Power source: AC electricity

Features: Visor lights up

Christian Ristow

These large-scale, kinetic sculptures are inspired by the work of Jean Tinguely and H.R. Giger and are intended to express the tension between man and machine. Ristow stages shows in which his sculptures smash, try to grab, and bite ferociously with powerful teeth. Most are animated by mechanical and electrical means, and pulse, throb, oscillate, and light up. According to Ristow, his robots aim to "fight each other and destroy designated sacrificial targets."

Specification: Manipulatrix

Date of manufacture: 2002

Country of origin: United States

Height: 5 ft 6 in (1.72 m)

Power source: Two-cylinder inline 17 horsepower gasoline engine

Features: Moves forward, grabs target objects, pulls objects into jaws and crushes them

Mandible
has chains to pull in victims

Hydraulic pistons
open the jaws

Platform for the operator is at the back of the robot

Tractor treads
move the robot forward

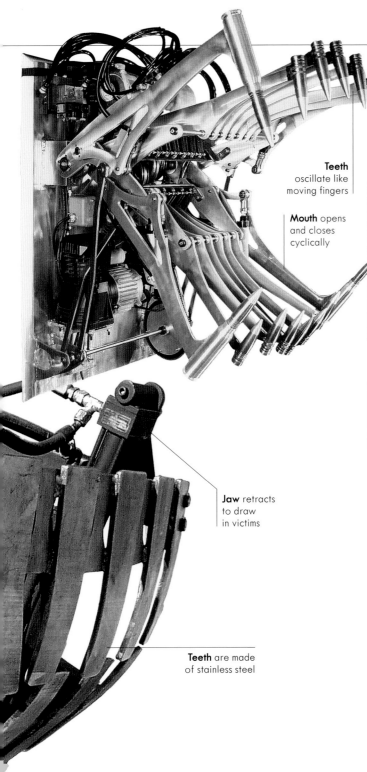

◄ Voracious Mouth

This air-compressor-driven robot mouth has inner and outer teeth that can open and close at different speeds. At the end of eight cycles, the mouth slams shut dramatically. The teeth are made from decommissioned 30-mm anti-tank practice rounds.

Teeth oscillate like moving fingers

Mouth opens and closes cyclically

Specification: Voracious Mouth
Date of manufacture: 2003
Country of origin: United States
Height: 3 ft 6 in (1.12 m)
Power source: Electric gear motor and onboard air compressor
Features: Teeth open and close, mouth slams shut

Specification: Impatient Hand
Date of manufacture: 2003
Country of origin: United States
Height: 18 in (45.7 cm)
Power source: Electric gear motor
Features: Taps on a metal surface using a rotating cam mechanism

Jaw retracts to draw in victims

Teeth are made of stainless steel

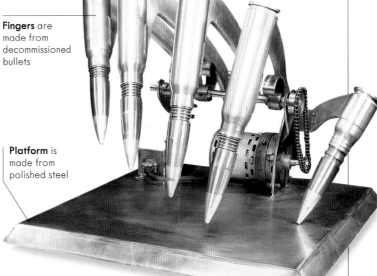

Fingers are made from decommissioned bullets

Platform is made from polished steel

◄ Manipulatrix

This gigantic robot is made of steel, stainless steel, and aluminum and is powered by gasoline, electricity, and hydraulics. Along the sides of the jaw are a pair of "mandibles," armed with spiked, rotating chains. These fearsome chains can grab objects and drag them into the waiting jaws.

▲ Impatient Hand

Through the use of a cam-style mechanism, the sculptor, Christian Ristow, has produced what he describes in his own words as a "rhythmic, cyclical tapping motion that resembles the human action associated with waiting." In contrast with many of Ristow's machines, this was designed to be displayed in an art-gallery setting.

Lawrence Northey

This Canadian sculptor has an unusual way of finding inspiration for his highly polished and whimsical creations. He begins by writing a science-fiction story, then bases his sculptures on his characters. The stories are inspired by Northey's belief that what we have been told about the past is in direct contrast with what is truly known. The robots' gestures and expressions add an emotional element to his art. Some of his sculptures move and make sounds.

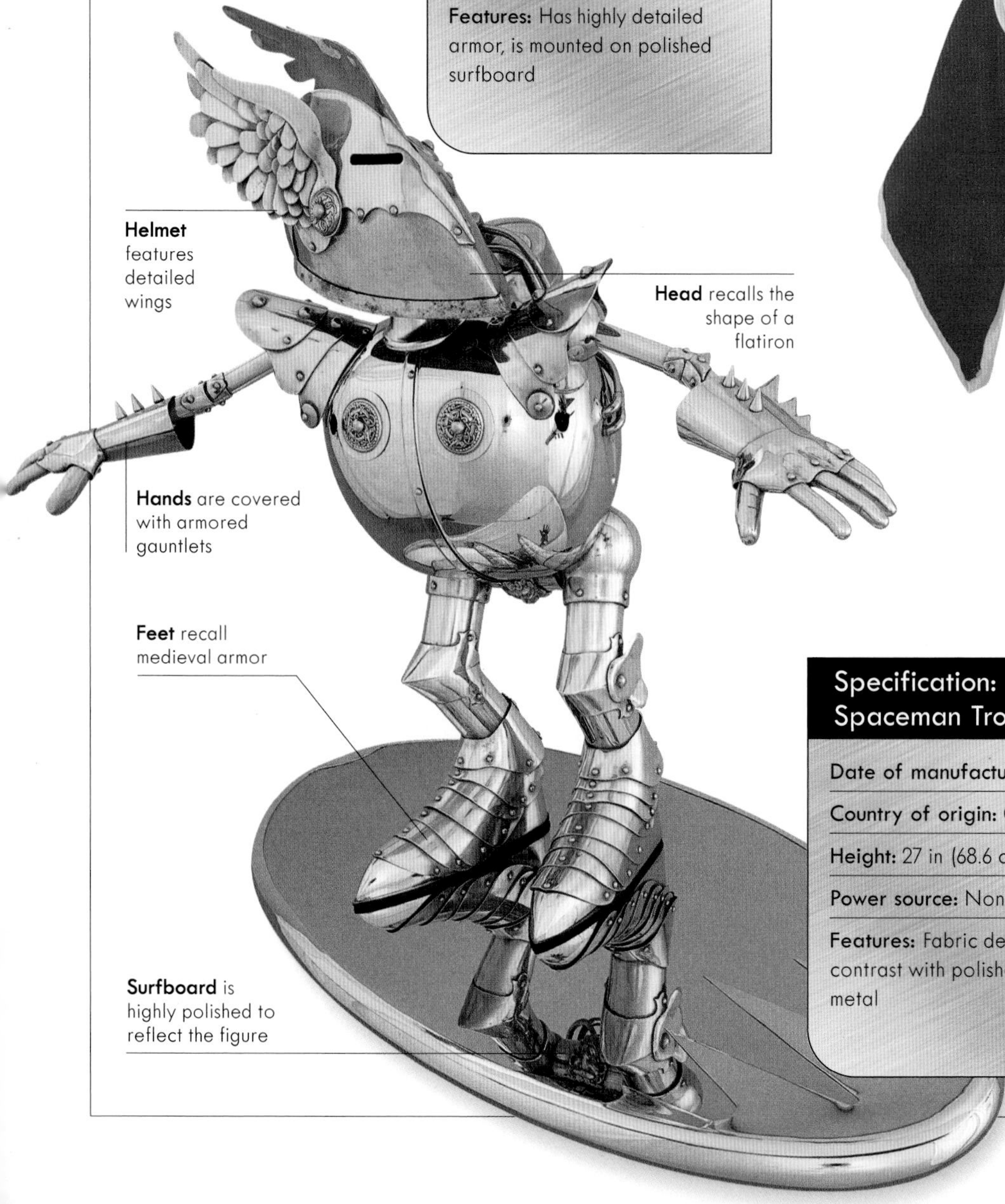

▶ Spaceman Troy

A childhood memory of a Hopi dancer is transformed into the artist's idea of a twenty-first century Kachina doll. Echoes of the knights of the Round Table and *Star Trek* paraphernalia influence the figure.

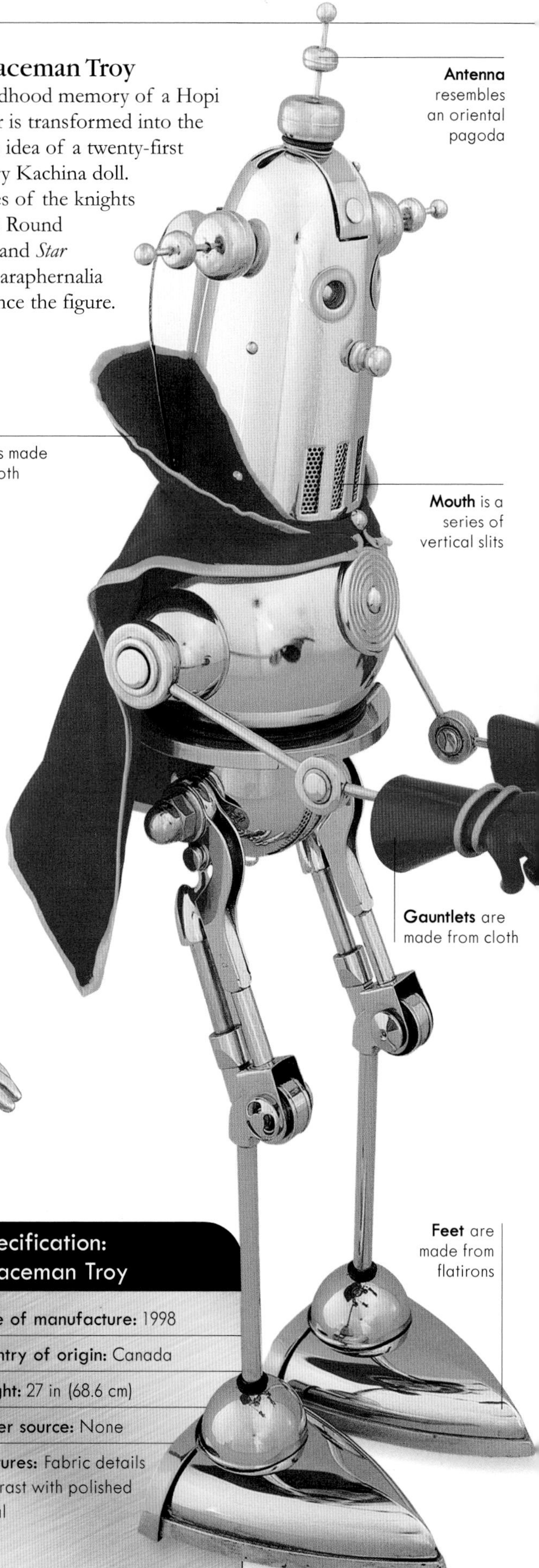

Antenna resembles an oriental pagoda

▼ Chantecler Eldorado

Created as a 3-D illustration for Northey's story *Wired City*, Chantecler Eldorado is shown here as the Game Master for the Royal Court.

Specification: Chantecler Eldorado

Date of manufacture:	1998
Country of origin:	Canada
Height:	28 in (71 cm)
Power source:	None
Features:	Has highly detailed armor, is mounted on polished surfboard

Cape is made from cloth

Mouth is a series of vertical slits

Helmet features detailed wings

Head recalls the shape of a flatiron

Hands are covered with armored gauntlets

Gauntlets are made from cloth

Feet recall medieval armor

Feet are made from flatirons

Specification: Spaceman Troy

Date of manufacture:	1998
Country of origin:	Canada
Height:	27 in (68.6 cm)
Power source:	None
Features:	Fabric details contrast with polished metal

Surfboard is highly polished to reflect the figure

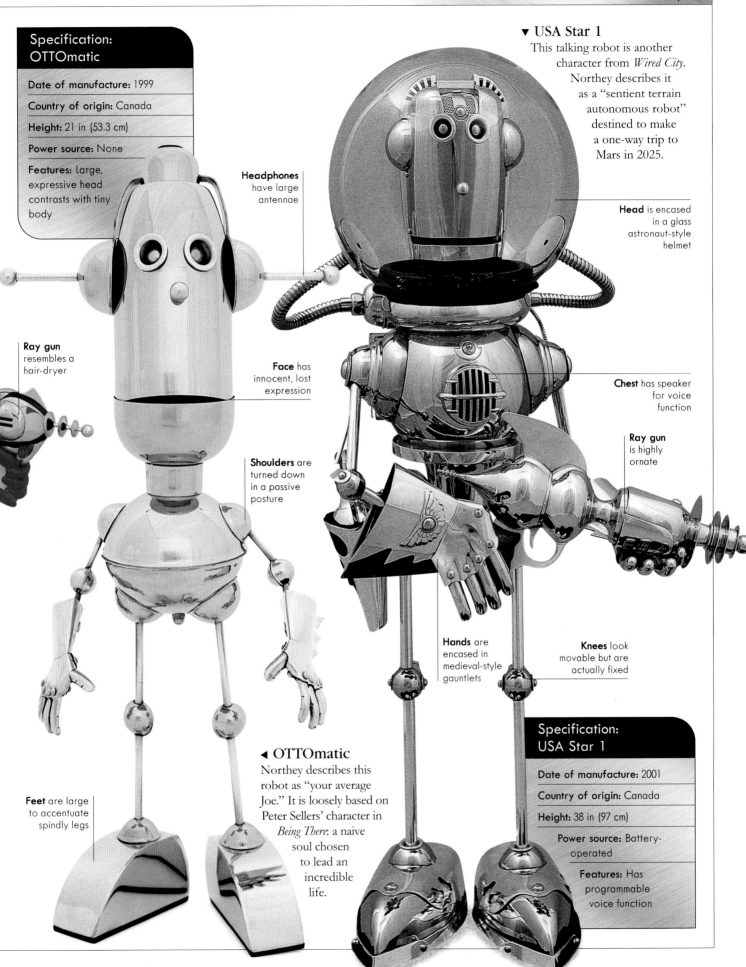

Specification: OTTOmatic

Date of manufacture: 1999

Country of origin: Canada

Height: 21 in (53.3 cm)

Power source: None

Features: Large, expressive head contrasts with tiny body

Ray gun resembles a hair-dryer

Feet are large to accentuate spindly legs

Headphones have large antennae

Face has innocent, lost expression

Shoulders are turned down in a passive posture

◄ OTTOmatic
Northey describes this robot as "your average Joe." It is loosely based on Peter Sellers' character in *Being There*: a naive soul chosen to lead an incredible life.

▼ USA Star 1
This talking robot is another character from *Wired City*. Northey describes it as a "sentient terrain autonomous robot" destined to make a one-way trip to Mars in 2025.

Head is encased in a glass astronaut-style helmet

Chest has speaker for voice function

Ray gun is highly ornate

Hands are encased in medieval-style gauntlets

Knees look movable but are actually fixed

Specification: USA Star 1

Date of manufacture: 2001

Country of origin: Canada

Height: 38 in (97 cm)

Power source: Battery-operated

Features: Has programmable voice function

Soccer robots

ALL AROUND THE WORLD robots are competing in contests that demonstrate an extraordinary variety of skills. From Sumo to flying, robots are becoming part of the sports world—they even have their own robot version of the Olympic Games, ROBOlympics. Soccer is the most popular sport for robots. All manner of robots designed by amateur enthusiasts, organizations, and universities are lining up for the kickoff and are aiming to be top of the cyber league.

▲ **Menacing scorer**
A Terminator-style competitor squares up to the goal mouth for a penalty shot in FIRA's individual humanoid league, HuroSot.

Human dominance on the soccer field could soon be under threat, now that robots are limbering up on the touchline. There are hundreds of regional robot soccer contests, and even tabletop "botball" competitions. However, two main tournaments dominate: RoboCup, and the FIRA

▲ **Ball for a head**
HanSaRam's ball head should beware of high-kicking rivals.

World Cup. RoboCup originated in Japan in 1993, and has held Open contests all over the world since 1997. Its rival, the FIRA (Federation of International Robot-soccer Associations) World Cup, began in Korea in 1996, and it too has spread worldwide.

The nearest human comparison is perhaps the UK's Football Association (FA) Cup, since even a small team can play a major-league opponent in both robot and human contests. The ultimate aim for the organizers of RoboCup is a playoff in 2050 between humanoid robots and the world champion human soccer team.

Although soccer contests are intended to be entertainment, there is a serious side: pushing back the boundaries of robots' skills. RoboCup holds annual symposiums for entrants and publishes research, while FIRA encourages entries from robotics undergraduates.

◀ **Fancy footwork**
Robo Erectus attempts to live up to its name in RoboCup 2003, playing for Singapore Polytechnic. It can walk, turn, crouch, and score a goal, all without falling over.

▲ **ASIMO'S soccer-player counterpart**
A cube-headed, modified version of Honda's ASIMO runs onto the field in full soccer gear. It featured in RoboCup 2003, representing Honda International Technical School, Japan.

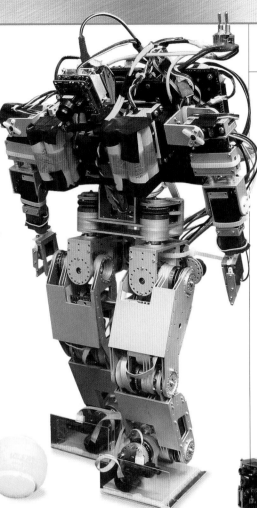

▶ **Sure shot**
Hiro, from the University of Manitoba, Canada, uses its complex sensor array to line up a shot in FIRA's HuroSot league.

Robotics united

Robot builders face major challenges: each player must be able to communicate with other team members efficiently while in action, as well as control the ball, avoid obstructions—and attempt to score goals.

FIRA's contests were originally solely between color-coded robots called MiroSots. These are hand-sized and cuboid, but very sophisticated. They play with an orange ball that is easily detected by their sensors. The contest has now broadened to include NaroSots—robots less than 2 in (5 cm) tall— and SimuroSots, which are software robot simulations, similar to computer games.

However, not surprisingly, by far the greatest publicity is given to the humanoid leagues in both of the rival tournaments. FIRA's humanoid robot (HuroSot) contest opens the field to a whole host of specially built bipeds that aim one day to match the skills of stars such as Honda's ASIMO. A version of ASIMO played in the rival RoboCup 2003 Humanoid tournament, while teams of AIBO dogs represented Sony in RoboCup's four-legged competition.

▶ **Ball in focus**
Osaka University's Senchans has its camera eye on the easily-identified orange ball in the RoboCup humanoid league.

Penalty kicks

In FIRA's HuroSot league, the competition is more limited than RoboCup's contest, since it is not open to the "corporate" humanoid robots. Participating teams are currently only expected to field single players that can walk steadily, avoid stationary obstacles, and take penalty shots—all under the remote guidance of a human trainer. But with RoboCup's field of star players, and its proposed robot-versus-human battle scheduled for 2050, it has its eye on an outright win, not a penalty shoot-out.

Nonhumanoid players

Robot soccer is mostly played between nonhumanoid teams of cube-shaped players, although their two-legged relatives tend to steal the limelight.

Sensors track the position of the ball

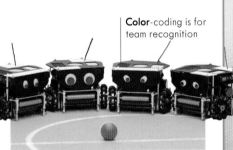

▲ **Cuboid contestants**
FIRA's MiroSot World Cup is between these small sensor-packed, color-coded bots fighting for possession of the ball. It remains the main FIRA event.

Color-coding is for team recognition

▲ **Playing as a team**
In RoboCup, small-sized League players from Cornell University work with a shared team intelligence.

Player hides the ball from its opponent

Colors identify team members

▲ **Dogged determination**
Sony's team of modified AIBOS compete in the Four-Legged soccer competition at RoboCup. Distance sensors enable AIBO to judge distances and avoid obstacles.

BattleBots

The international phenomenon of BattleBots is a televised fight to the death between home-made robot gladiators. Arming their creations with weapons that pulverize, slash, and rip, amateur teams build robots to meet in a bloodbath of engine oil and metal on metal, in a brutal, hazard-filled arena. Only the strongest will survive.

BattleBots Logo

The competition

To compete in BattleBots, teams have to build their robots within the rigorous tournament specifications set out by BattleBots Inc. Judges score robots' performance in three categories: aggression (the frequency, severity, boldness, and effectiveness of attacks); damage (to opponents' functionality, effectiveness, and defensibility); and strategy (a controlled offensive plan, or a defensive one that protects the competitor's weaknesses against a specific opponent).

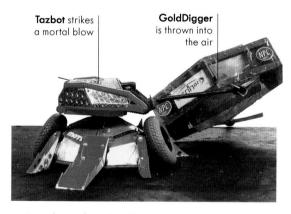

Tazbot strikes a mortal blow

GoldDigger is thrown into the air

▲ Duel to the death
Tazbot and GoldDigger clash in a ferocious bout. The ferrous, laser-levelled arena is strewn with hazards to catch even the most daring robot unaware. The hazards are as terrifyingly named as the robots themselves: Kill Saws; Pulverizers; Hell Raisers; Ram Rods; the Spike Strip; and the Vortex.

Diesector

Super-heavyweight champion Diesector, built by Donald Hutson, has been in the top-three rankings since its debut, and has bashed its way to reigning supreme. A Super-heavyweight BattleBot has to weigh more than 100kg (220lb), but less than 154kg (340lb) in its battle-ready configuration.

▼ Killer robot
The formidable design and colour scheme of this mutant, industrial ClampBot sends out a clear message: danger!

Mallets can crush opponents

Specification: Diesector

World ranking: 1

Country of origin: USA

Power Source: Battery

Primary weapons: Jaws and battle mallets

Capabilities:
Uses battle mallets to pound opponents

BioHazard

Built by Carlo Bertocchini, BioHazard is the most successful Heavyweight champion ever, almost always ranking number one. To qualify as a Heavyweight, a BattleBot has to weigh more than 54kg (120lb), but less than 100kg (220lb) in its battle-ready configuration.

▼ Bottom-heavy
A low centre of gravity makes BioHazard difficult to beat, and its powerful central arm is a formidable weapon in the arena.

Weapon is in the form of a single arm

Specification: BioHazard

World ranking: 1

Country of origin: USA

Power source: Battery

Primary weapon:
Lifting arm

Capabilities: Uses a dual linear actuated arm to hurl opponents

Base is broad to provide stability

Dr. Inferno Jr.

A BattleBots superstar, Dr. Inferno Jr. has sawed up the opposition to grab its top-three Lightweight ranking. Built by Dr. Jason Bardis as the progeny of Dr. Inferno, Junior combines powerful weaponry with a toy-like face. Lightweight BattleBots weigh between 30 lb (14 kg) and 60 lb (27 kg), while Middleweights weigh from 60 lb (27 kg) to 120 lb (54 kg).

▼ **Strong arm**
Dr. Inferno Jr. has deadly fighting arms to which can be attached electric saws and drills, bludgeons, spikes, and boxing gloves.

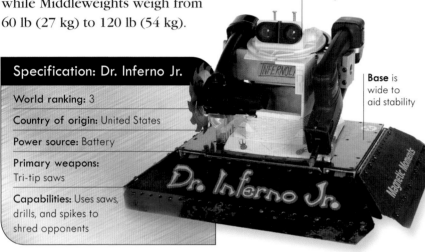

Face makes robot look deceptively friendly

Base is wide to aid stability

Specification: Dr. Inferno Jr.

World ranking:	3
Country of origin:	United States
Power source:	Battery
Primary weapons:	Tri-tip saws
Capabilities:	Uses saws, drills, and spikes to shred opponents

Nightmare

One of a series of beautifully designed robots by Jim Smentowski of Team Nightmare, this robot has consistently been voted the most aggressive in the Heavyweight class by other teams. Nightmare is the dark horse of the show, always ranking highly, but threatening to do better. Team Nightmare's other robots include Backlash, Micro-Nightmare, Turtle, Shazbot, and Locust of the Swarm.

▼ **Spinning terror**
Nightmare's main weapon is its rotary cutting and bludgeoning disc. The fearsome weapon is capable of delivering a blow as powerful as that of a freight train.

Teeth are attached to a rotary-saw-style disc

Specification: Nightmare

World ranking:	6
Country of origin:	U.S.
Power source:	Battery
Primary weapon:	Rotary bludgeoning device
Capabilities:	Uses spinning disc to slice up and smash opponents

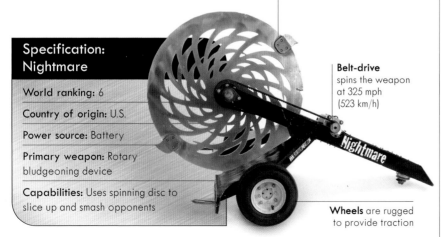

Belt-drive spins the weapon at 325 mph (523 km/h)

Wheels are rugged to provide traction

An arsenal of weaponry

The BattleBots arena welcomes all entrants, as long as they satisfy the rigorous entry requirements. Builders take this as an opportunity to showcase some brilliant design work.

▲ **Mauler**
The beautifully decorated Mauler is a Heavyweight Bot, from the team of Commander Tilford and Crew. One of the first full body-spinning BattleBots, its world ranking is number 10.

▶ **Phere**
This unusual looking Super-heavyweight BattleBot is ranked at number 26. Built by Gaylan Douglas, its primary weapons are the spinning dome with corkscrew blades.

Blade is serrated for extra impact

▲ **The Master**
Built by Mark Setrakian, this Middleweight Bot has a world ranking of 42. Its primary weapons are a gas-powered cutoff saw and actuated lifting arm.

Wedge can slide beneath opponents to unbalance them

▲ **Bad Attitude**
This speedy Middleweight Bot was built by Thomas Petruccelli. Its primary weapon is the wedge, and it was built for maximum durability, power-to-weight ratio, and speed.

GALLERY: Robots in Art & Entertainment

The robot has long held a fascination for moviemakers and artists, with fictional robotic characters being almost too numerous to list. The twentieth-century love of science fiction meant that robots increasingly took center stage. In a fantasy world of endless possibilities, the only limits to the robots' capabilities lay in the boundaries of their creators' imaginations. Although these robots may require technology beyond the wildest dreams of scientists to turn them into reality, they continue to inspire us to dream about what could be.

Maître de Mystere
A huge, mechanical robot presents yet another challenge to the chained escape artist Houdini in this poster for his movie *Maître de Mystere*. The movie was produced in 1920 by the French studio Pathé.

Undersea Kingdom
This is a promotional poster for the 1936 sci-fi serial, made by Republic Studio. It tells the story of Ray "Crash" Corrigan, who discovers Atlantis and then has to battle the evil ruler, Unga Khan, and his army of robot soldiers.

Dr. Satan's Robot
In this remake of a 1940s movie series, the hero, Copperhead, tries to stop a criminal and his robot from taking over the world.
- **Date:** 1966 ■ **Country of origin:** United States ■ **Height:** 6 ft 5 in (2 m) ■ **Studio:** Republic

The Spectre of Suicide Swamp
E.K. Jarvis's story about a robot that created terror in a swamp is illustrated by Walter Popp in a 1952 edition of *Fantastic Adventures* magazine, published by Fantastic Adventures/Ziff Davis.

Menace of the Metal Men
A magazine cover by S.R. Dirgin features a diabolical robot rising from the mad scientist's operating table from *Menace of the Metal Men*.
- **Date:** 1938/39 ■ **Country of origin:** United States/UK ■ **Publisher:** Fantasy

Mother Riley Meets the Vampire
Old Mother Riley looks up in terror at a uranium-powered slave robot, one of a group of robots created by the mad scientist Von Hoosen, (Bela Lugosi). The movie was also known as *The Vampire and the Robot*.
- **Date:** 1951
- **Country of origin:** UK ■ **Height:** 6 ft (1.8 m) ■ **Studio:** Renown Pictures

C-3PO and R2-D2

An early version by illustrator Ralph McQuarrie of the robots C-3PO and R2-D2 for the 1977 movie *Star Wars*. Working closely with George Lucas, McQuarrie helped to conceptualize many of the movie's characters. The facial features of his version of C-3PO are quite different from those of the robot in the movie.

Total Recall

In one of this movie's most memorable scenes, Johnny Cab, the robot taxi driver, keeps functioning even after the taxi crashes.
■ Date: 1990 ■ Country of origin: United States
■ Height: 3 ft (90 cm) ■ Studio: Carolco/Tristar

Cyborg Cop

Captured by a drug lord after a failed undercover operation, DEA agent Ryan is turned into a cyborg.
■ Date: 1993 ■ Country of origin: U.S.
■ Height: 6 ft (1.8 m) ■ Studio: Nu World

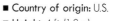

Rocket Man

This sculpture by Clayton Bailey imaginatively combines a robot and a rocket poised for takeoff.
■ Date: 1995
■ Country of origin: U.S.
■ Height: 4 ft (1.2 m)

Zom-01

This powerful fighting robot is from the *Zentrix* computer animation series. Zom-01 has powerful arms and super armor.
■ Date: 2002 ■ Country of origin: Hong Kong
■ Studio: Imagi

Tokyo Teen

Sculptor Lawrence Northey gives us his interpretation of a knock-kneed teen with flatiron feet, hoola hoop, and headphones.
■ Date: 2003 ■ Country of origin: Canada
■ Height: 32 in (81 cm)

Robot

This 2001 picture by American artist David Rose is typical of his retro-style paintings of robots. This one is acrylic painted on cardboard and represents a tinplate robot toy in a box, as if for sale.

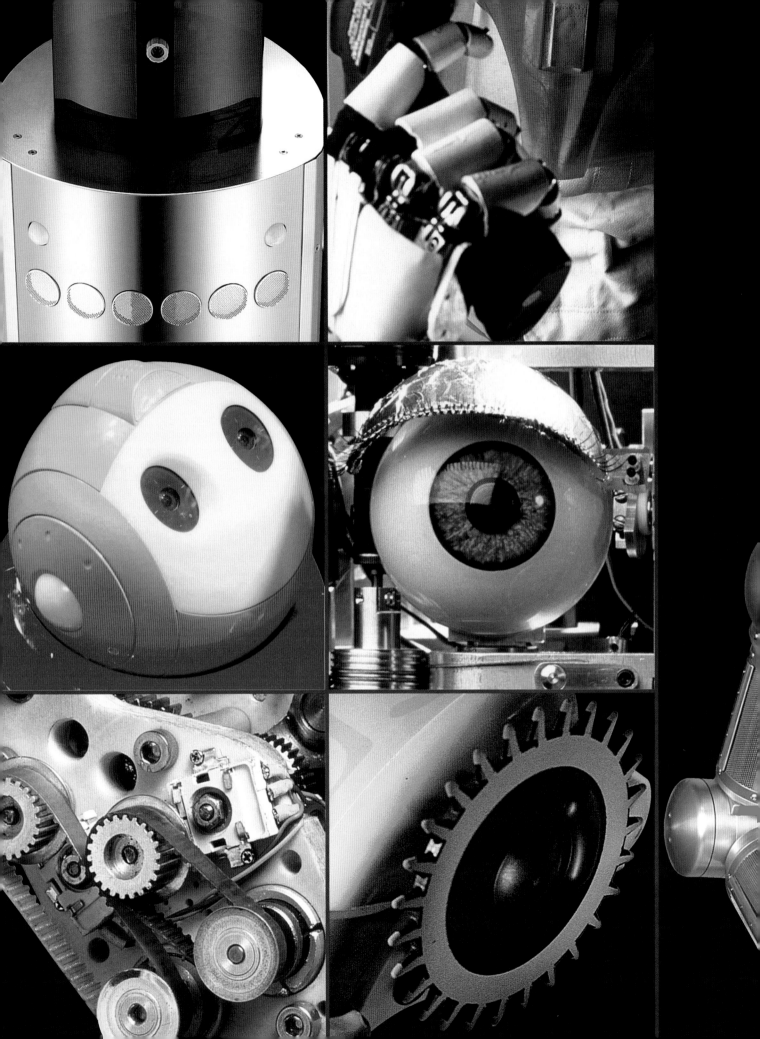

The New Generation

Living with Robots

The da Vinci surgical hand

SCIENCE-FICTION WRITER RAY BRADBURY once said that he wrote about things to prevent them from happening. Just a few decades after the first fictional incarnations, the robots really are coming. Although today's robots are designed to help us at home and at work, and to entertain us, soon we may face some of the dilemmas that science-fiction stories have warned us about.

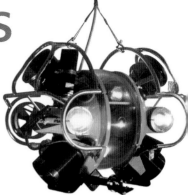

Hyball, a fully autonomous miniature submarine

A new generation of robots that can walk and talk like humans is already here. More robots will take their places in our living rooms, factories, and public spaces, each one learning the lessons of the previous generation of robots—and learning them at Internet speed. There are already commercially available robots that can tend to the sick, provide remote medical advice, or simply act as companions. Robots such as Wakamaru, Dr. Robot, Nuvo, and QRIO can monitor home security, surf the Internet when prompted by voice commands, and email video images to a cell phone. Some robots can sing, dance, walk, and run just like their human counterparts; a few, such as Kawada Systems' HRP-2, can even pick themselves up off the floor if they fall.

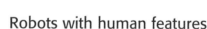

Nursebot Pearl, designed to help the elderly in their homes

Forging ahead

Robots are also making the first tentative steps of the journey to new frontiers. Some have already gone before us to Mars, paving the way, perhaps, for future human colonization. In the near future, robots may go farther into outer space, but also venture deep into inner space, since miniaturization could enable the creation of nano-robots—robots that are small enough to move around in the bloodstream.

Although universities and research centers are at the forefront of robotic research, the military and security services are also involved in robot development. Robots now patrol airports and other sensitive locations, confronting intruders and tackling security threats.

Robots with human features

One thing is certain: robots will progress as only robots can: becoming smaller and faster, and traveling farther. They are also becoming more like us. Robot faces can smile, frown, register emotions, and communicate

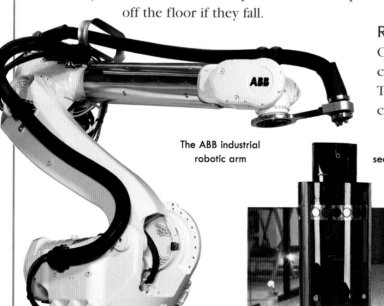

The ABB industrial robotic arm

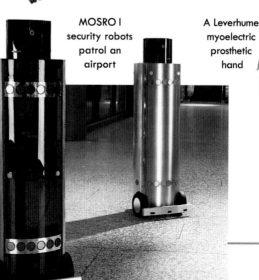

MOSRO I security robots patrol an airport

A Leverhume myoelectric prosthetic hand

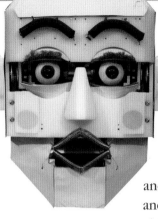

The WE-4 robot face, which registers emotions

with us, and robot hands are flexible and accurate enough to perform delicate operations. Such advances are not mere gimmicks: they allow us to feel more comfortable working with and around humanoid machines, and they complement our human abilities with their machine precision and reliability.

Skill and intelligence

Even the current generation of robot toys points the way ahead to an exciting future. The gradual, predicted switch from machinelike robots to humanoids is an outward sign of a large-scale inner change. We have become used to automated processes, and to low-cost, high-specification computing. If the two are put together, we have a clear vision for the development of robots as humanlike tools with high levels of skill and computing intelligence. Today, we have the Internet, mobile digital communications, and open-source operating systems. All of these advances allow people to collaborate on projects, sharing their skills and knowledge wherever they are in the world. Soon we will have the Grid, a far more advanced network of high-specification computers, plus fast, mobile video communications. All of this suggests an exciting future for robots and robotics based on this technology.

HOAP-1 miniature humanoid robot

Machine-made machines

Currently, robots are still designed by humans, programmed by humans, and assigned work by humans. However, the future of robots is shared with that of computer technologies. One day, perhaps, machines will begin programming other machines. If

that happens, robots may begin to develop more quickly than humans can, and this could threaten our supremacy. Until that giant evolutionary leap, robots will predominantly be slave-like workers, obeying the origin of their name.

Robots' development can certainly give us a new perspective on the place of advanced technology in

An Ifbot personal robot, which can recognize faces

our professional lives. However, the robotics revolution is only partly to do with industrial applications; it is also related to the spread of lifestyle gadgets in the home. Humans are more comfortable with technology than ever before, so robots' time may finally have come.

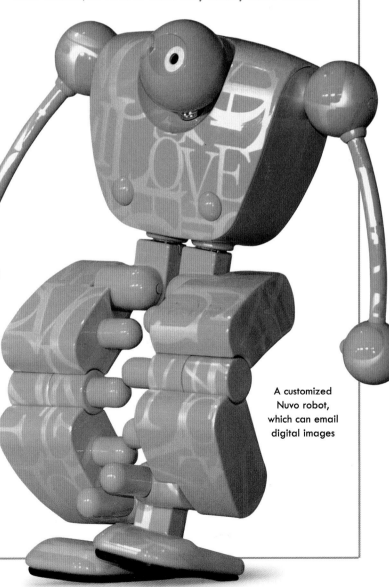

A customized Nuvo robot, which can email digital images

Little Helpers

Opinions differ on the role of personal robots, whether they should assist, entertain, or do both. PaPeRo and Maron are, to a degree, robots of the third kind – they do not do the housework, but they are responsive and far more intelligent than their ancestors, such as RB5X and Topo of 20 years ago.

Specification: PaPeRo	
First manufactured: 2004	
Country of origin: Japan	
Manufacturer: NEC	
Height: 15 in (38.5 cm)	
Power source: Battery-operated	
Intelligence: Onboard microprocessor	
Capabilities: Recognizes speech, talks, moves and responds, controls household devices	

PaPeRo

An innocent-looking exterior belies PaPeRo's intelligence, giving it twice the appeal by hiding cutting-edge technology behind a friendly face. This robot recognizes 650 phrases and 3,000 words. It can also remotely control a TV and other household devices, and is Internet-enabled via a wireless connection. But PaPeRo has first and foremost been designed to put people at ease: its sensors can distinguish between a pat and a stroke, and it can even sense when it is being lifted up.

Eyes contain twin digital cameras for stereo imaging and positioning

Backpack makes the robot appear childlike

◄ **Bag on its back**
PaPeRo comes in a range of colors and has various accessories, such as a backpack. As well as maximizing the robot's cute appeal, the backpack can also be used to carry spare parts.

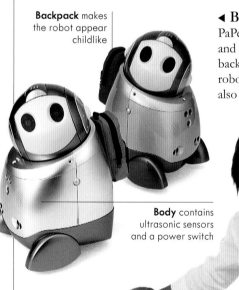

Body contains ultrasonic sensors and a power switch

▲ **Transparent technology**
This transparent version strips away the friendly exterior to reveal the electronic workings within. Inside are two digital CCD cameras and three microphones, one of which is omnidirectional for better speech recognition, plus a host of other sensors and mechanics.

Body contains motion and gyroscopic sensors

Ear contains LEDs

Sensor in the head can detect touch

◄ **Eye contact**
PaPeRo is known for its ability to communicate with more than one person at a time. It can turn its head to face each person and make eye contact. The eyes blink when it senses people, and turn orange when it recognizes them.

Body contains electronics, batteries, and speakers

▼ **Cute but smart**

PaPeRo comes in a range of colors. Some people might see in its shape and expression a resemblance to a cartoon character, or a Russian doll, and such responses are precisely what today's robot designers want to evoke. Manufacturers have realized that many people feel uncomfortable with the idea of artificial beings, so the challenge is to build robots that we feel protective toward, rather than threatened by.

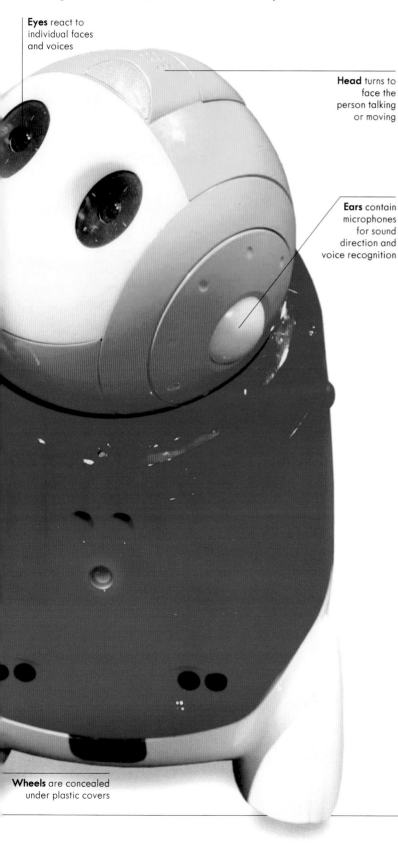

Eyes react to individual faces and voices

Head turns to face the person talking or moving

Ears contain microphones for sound direction and voice recognition

Wheels are concealed under plastic covers

Maron

This nonhumanoid robot is designed to be a guard rather than an interactive companion. It can be controlled remotely via a cell phone and used as a camera to search premises. It also sounds an alarm if it detects an intruder. Maron is able to turn on a VCR, a dishwasher, and other appliances.

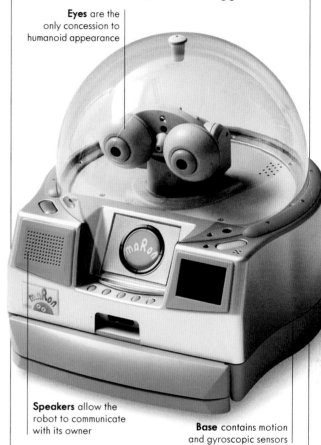

Eyes are the only concession to humanoid appearance

Speakers allow the robot to communicate with its owner

Base contains motion and gyroscopic sensors

▲ **Wheeling along**

Maron has no legs or arms and rolls along on two wheels. It is equipped with a 12-hour battery charge and an AC adapter.

Specification: Maron	
First manufactured: 2002	
Country of origin: Japan	
Manufacturer: Fujitsu	
Height: 12½ in (32 cm)	
Power source: Battery-operated	
Intelligence: Onboard microprocessor	
Capabilities: Talks, relays photographs, navigates, operates appliances, sounds alarm, calls preset telephone numbers	

Sico Millennia

This is the most advanced member of the Sico robot family. Its elegant design conceals sophisticated technology that allows it to perform a range of complex tasks. Millennia has appeared on TV and taken part in a fashion shoot for *Vogue* magazine. It can communicate in seven languages and is the only robot member of the Screen Actors' Guild.

Celebrity robots

The Sico family are working entertainment robots, designed to appear at conferences or events in the role of mascot or host. They can be preprogrammed to deliver an entire staged and scripted show; or they can be operated live by remote control. Sico robots even have their own American Express cards and travel first class. They can be hired with a professional actor who operates them, or they can be custom-designed to meet a client's specifications.

Head rotates fully and moves up and down

Eyes contain onboard cameras, which allow the operator to view what the robot is "seeing"

Mouth lights up as the robot speaks through hidden internal speakers

Chest has a built-in message-delivery screen

Shoulder joint rotates on one axis only

Specification: Sico Millennia

First manufactured: 2002

Country of origin: United States

Manufacturer: International Robotics

Height: 6 ft (1.8 m)

Power source: Intelligent motors

Intelligence: Microprocessor, plus remote control

Capabilities: Moves, speaks, "sees" using onboard cameras, sings, interacts with people

Hand has a gentle open-and-close action

Waist has a sophisticated pivoting mechanism

Rubber bumper protects the base

▶ **Sophisticated functions**
Sico Millennia can fluidly rotate its waist, head, and body in opposite directions to perform a dance routine. It has built-in collision sensors, and can safely carry a person on its integral passenger platform.

Neck allows the head to move up, down, and around, and it can tilt

Base houses motors developed for aerospace applications

Passenger platform enables robot to offer rides and dances with guests

Elbows have intelligent motors, allowing limbs to move swiftly and intuitively

▶ **Style and practicality**
Sico Millennia has been designed to look sleek and elegant but still be hard-wearing enough to withstand frequent use and travel. Custom-designed skins can be created to suit individual customers' needs.

ASIMO

The Honda humanoid takes its name from that visionary of robotics, Isaac Asimov. The evolution of ASIMO began in 1986, through a succession of prototypes—P1, P2, and P3—that evolved from studies of human movement. Today, ASIMO certainly represents an Advanced Step in Innovative Mobility (for which it is the acronym). It can walk as well as the humans it is designed to assist—or perhaps replace.

Flexible friend

This endearing robot was designed to be child-sized, so that people would feel protective of it rather than threatened by it. ASIMO can stand on one foot, walk backwards, navigate stairs, dance, and perform simple tasks, such as switching on a light. It has 26 degrees of freedom, which means that it can move in 26 directions.

▼ Humanoid Prototype P1

Honda's first humanoid prototype translated earlier studies of human motion into a fully working model with sophisticated arm and leg movements. The robot can grip and carry an object with the aid of an external controller.

Head is bulky and nonhumanoid

Frame is unwieldy

▼ Autonomous P2

Honda's first fully autonomous humanoid, P2, has its core components, such as motor drives, computers, battery, and radio, located within its body. P2 can operate either automatically or through wireless assistance.

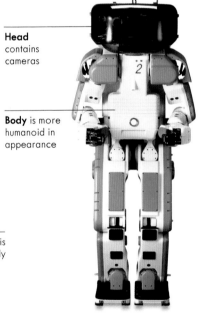

Head contains cameras

Body is more humanoid in appearance

▼ Elegant P3

Shorter and considerably lighter than P2, the humanoid robot P3 has evolved through miniaturization and redesign. The weight has been reduced using lighter metal and a more sophisticated distributed control system.

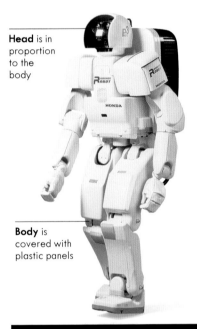

Head is in proportion to the body

Body is covered with plastic panels

Specification: P1
First manufactured: 1993
Country of origin: Japan
Manufacturer: Honda Motor Co. Ltd.
Height: 6 ft 3 in (1.9 m)
Power source: Battery-operated
Intelligence: Onboard computer and external controllers
Capabilities: Walks, carries objects, moves arms

Specification: P2
First manufactured: 1996
Country of origin: Japan
Manufacturer: Honda Motor Co. Ltd.
Height: 6 ft (1.8 m)
Power source: Battery-operated
Intelligence: Onboard computer and wireless assistance
Capabilities: Walks, climbs and descends stairs, pushes a cart

Specification: P3
First manufactured: 2000
Country of origin: Japan
Manufacturer: Honda Motor Co. Ltd.
Height: 5 ft 3 in (1.6 m)
Power source: Battery-operated—wireless Ethernet system
Intelligence: Back-mounted CPU
Capabilities: Walks, turns corners, carries objects, climbs up and down stairs

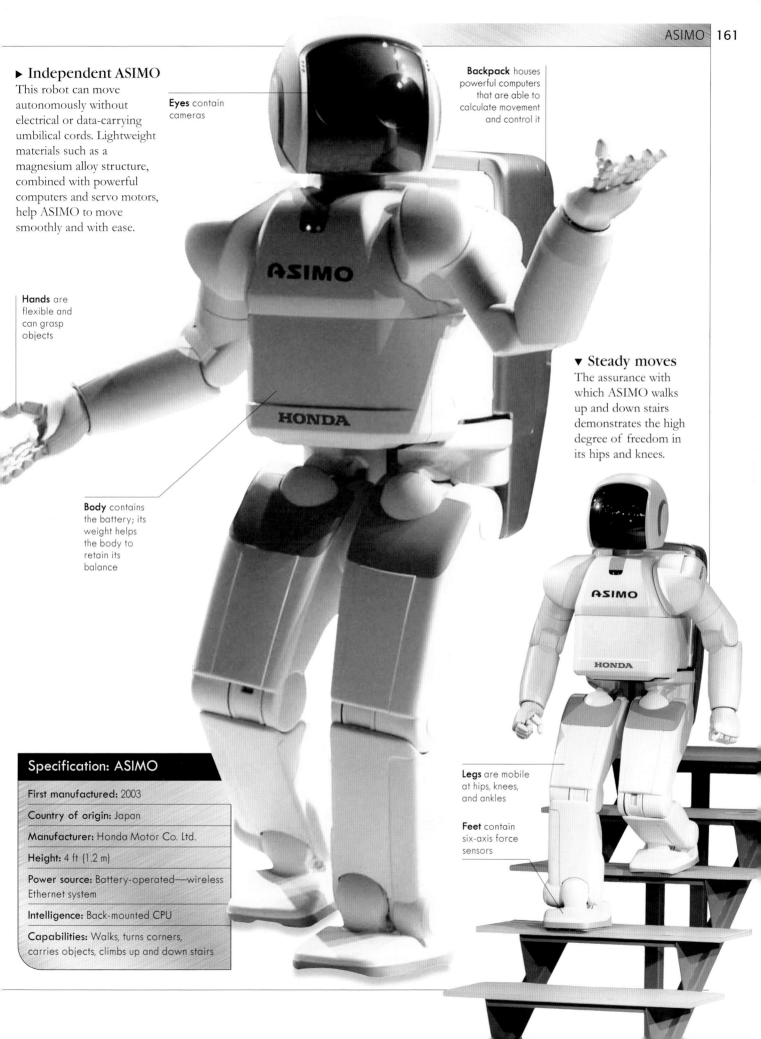

▶ **Independent ASIMO**
This robot can move autonomously without electrical or data-carrying umbilical cords. Lightweight materials such as a magnesium alloy structure, combined with powerful computers and servo motors, help ASIMO to move smoothly and with ease.

Eyes contain cameras

Backpack houses powerful computers that are able to calculate movement and control it

Hands are flexible and can grasp objects

Body contains the battery; its weight helps the body to retain its balance

▼ **Steady moves**
The assurance with which ASIMO walks up and down stairs demonstrates the high degree of freedom in its hips and knees.

Legs are mobile at hips, knees, and ankles

Feet contain six-axis force sensors

Specification: ASIMO

First manufactured: 2003

Country of origin: Japan

Manufacturer: Honda Motor Co. Ltd.

Height: 4 ft (1.2 m)

Power source: Battery-operated—wireless Ethernet system

Intelligence: Back-mounted CPU

Capabilities: Walks, turns corners, carries objects, climbs up and down stairs

HOAP Robots

Aimed at academics and researchers, the optimistically-named HOAP stands for Humanoid for Open Architecture Platform. Made by Fujitsu, HOAP-2 is programmed by computer via USB and, with its earlier version, HOAP-1, can be run on the Linux operating system, allowing users to develop their own programs for the robots. Despite its similarity, HOAP-2 is a major step forward from HOAP-1; both robots are renowned for their strength and agility.

HOAP-1

Using a gyroscope and accelerometers in its chest for balance, HOAP-1's Intel processor tracks the robot's center of gravity. Force sensors in the heels and toes fine-tune its balance and relay data on which leg is supporting the robot, triggering the other leg to lift. HOAP-1 has 20 degrees of freedom. It is battery-operated, but for sustained motion it needs a power supply. Fujitsu supplies software-development tools and a computer simulator with every robot.

Specification: HOAP-1

First manufactured:	2002
Country of origin:	Japan
Manufacturer:	Fujitsu
Height:	19 in (48 cm)
Power source:	Battery-operated
Intelligence:	Intel MMX Pentium chip
Capabilities:	Walks, climbs stairs, sits, hops, bends, twists at the hips

▼ **Walking miracle**
The HOAP was developed as a test platform for bipedal locomotion. It was the first attempt to market a humanoid robot that users could program themselves.

Head rotates and contains a camera

Upper arm is controlled by a local CPU

Backpack supplies power via exposed cabling

Hips have three degrees of freedom to aid balance

Knee is operated by a single microcontroller

Ankle can flex and twist for balance

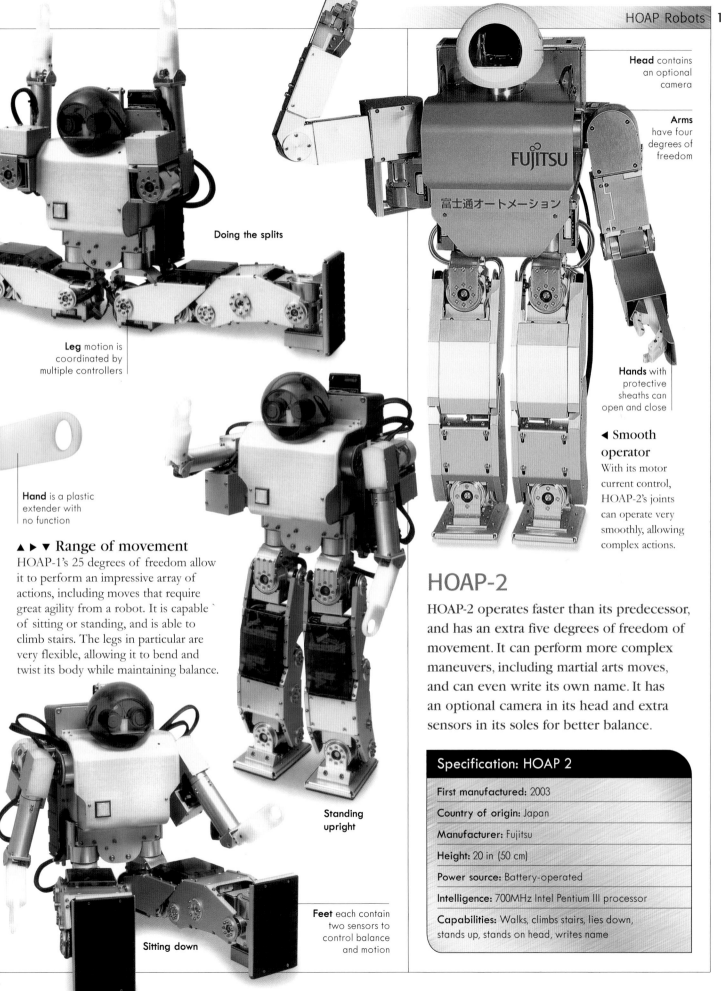

Doing the splits

Head contains an optional camera

Arms have four degrees of freedom

FUJITSU

富士通オートメーション

Leg motion is coordinated by multiple controllers

Hands with protective sheaths can open and close

◄ **Smooth operator**
With its motor current control, HOAP-2's joints can operate very smoothly, allowing complex actions.

Hand is a plastic extender with no function

▲ ► ▼ Range of movement
HOAP-1's 25 degrees of freedom allow it to perform an impressive array of actions, including moves that require great agility from a robot. It is capable ` of sitting or standing, and is able to climb stairs. The legs in particular are very flexible, allowing it to bend and twist its body while maintaining balance.

Standing upright

HOAP-2

HOAP-2 operates faster than its predecessor, and has an extra five degrees of freedom of movement. It can perform more complex maneuvers, including martial arts moves, and can even write its own name. It has an optional camera in its head and extra sensors in its soles for better balance.

Feet each contain two sensors to control balance and motion

Sitting down

Specification: HOAP 2
First manufactured: 2003
Country of origin: Japan
Manufacturer: Fujitsu
Height: 20 in (50 cm)
Power source: Battery-operated
Intelligence: 700MHz Intel Pentium III processor
Capabilities: Walks, climbs stairs, lies down, stands up, stands on head, writes name

Kawada Robots

These futuristic-looking robots may seem like life-sized Transformer toys, but the reality is very different: these are robots designed to work alongside human beings in a range of industrial and domestic settings. Kawada's HRP-2 and its HRP-2P prototype are in a class with Honda's ASIMO and Sony's playful QRIO, but they are much larger than both. Having gone through extensive research and development, these may be the strongest and most agile of all humanoid robots.

HRP-2P and HRP-2

HRP-2P is the prototype development platform for HRP-2. During its genesis, it shed a cumbersome backpack as its designers built in higher-density electronic components. HRP-2P has learned to walk down a narrow path and on uneven surfaces, and can pick itself up off the floor very elegantly if it falls.

▶ **Working robot**

HRP-2P was developed to be used to work in dangerous areas, care for the elderly or disabled, or even do domestic chores around the home. The robot's torso is operated by a vibration gyro and velocity sensor.

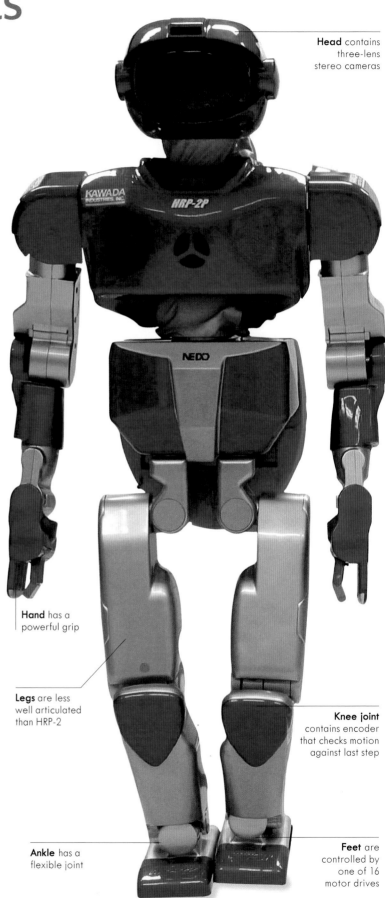

Head contains three-lens stereo cameras

Hand has a powerful grip

Legs are less well articulated than HRP-2

Knee joint contains encoder that checks motion against last step

Ankle has a flexible joint

Feet are controlled by one of 16 motor drives

Specification: HRP-2P

First manufactured: 2002

Country of origin: Japan

Manufacturer: Kawada Industries

Height: 5 ft (1.5 m)

Power source: NiMH DC 48-volt battery

Intelligence: ART-Linux powered with CPU and actuator drive system

Capabilities: Walks, gets up from prone position, walks on uneven surfaces, grasps objects

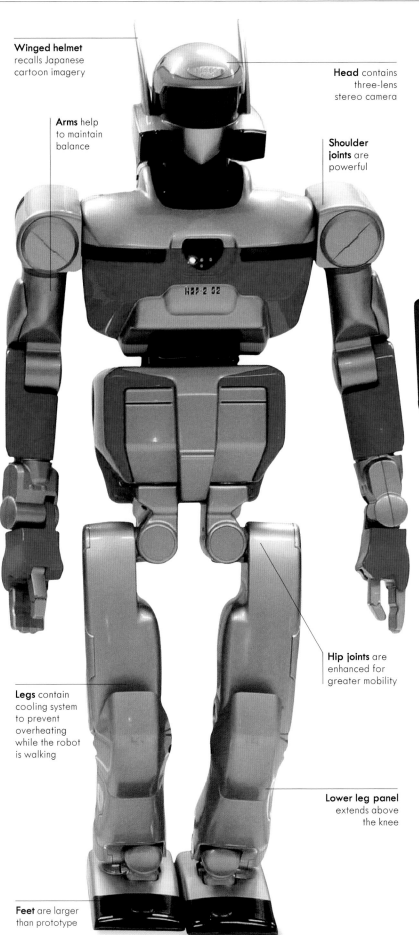

Winged helmet recalls Japanese cartoon imagery

Arms help to maintain balance

Legs contain cooling system to prevent overheating while the robot is walking

Feet are larger than prototype

Head contains three-lens stereo camera

Shoulder joints are powerful

Hip joints are enhanced for greater mobility

Lower leg panel extends above the knee

Legs are positioned under the center of the body for balance

▲ Lifelike movement

Getting up off the floor is not easy for a robot, but HRP-2 achieves it in a very lifelike way. Its enhanced hip and pelvic joints allow a great range of movement. The robot can walk, crouch, and get up off the floor as swiftly and smoothly as a human.

◄ Elegant humanoid

According to its developers, the HRP-2 is the final result of their humanoid studies. Yaskawa Electric provided the initial design concept—the winged-ear look was created by mechanical animation designer Yutaka Izubuchi—and the Vision Research Group and Shimizu Corporation developed the robot's vision system. HRP-2's 30 degrees of movement capability has been enhanced by significantly improving the joints used for the prototype.

Specification: HRP-2
First manufactured: 2003
Country of origin: Japan
Manufacturer: Kawada Industries
Height: 5 ft (1.5 m)
Power source: NiMH DC 48-volt battery
Intelligence: ART-Linux powered with CPU and actuator drive system
Capabilities: Walks, gets up from prone position, walks on uneven surfaces, grasps objects

QRIO

The product of cutting-edge artificial intelligence and dynamics technology, Sony's QRIO emerges from its previous prototype incarnations to be one of the most talented stars among small-scale humanoid robots. QRIO—pronounced "curio," and standing for "quest for curiosity" in Japanese—was designed by Sony to "live with you, make life fun, and make you happy,"

Specification: QRIO	
First manufactured: 2003	
Country of origin: Japan	
Manufacturer: Sony	
Height: 2 ft (61 cm)	
Power source: Battery-operated	
Intelligence: Microprocessors and PC link	
Capabilities: Walks, talks, runs, dances, plays ball games, surfs the Web, recognizes voices and faces, can differentiate between sounds	

Sony's dream robot

This talented robot can walk on uneven surfaces and, if it falls, pick itself up and check for damage. Inside your home, it will talk to you, complete with calming body language, once it has recognized your face and voice. It can also connect to the Internet and may soon be surfing the Web for you.

▼ Playing ball

An earlier version of QRIO, SDR-3X, plays soccer. The Intelligent Servo Actuator system allows the robot to perform amazingly complex movements in an autonomous mode of operation. Playing ball demonstrates the robot's coordination, agility, and fluidity of movement as well as its balance.

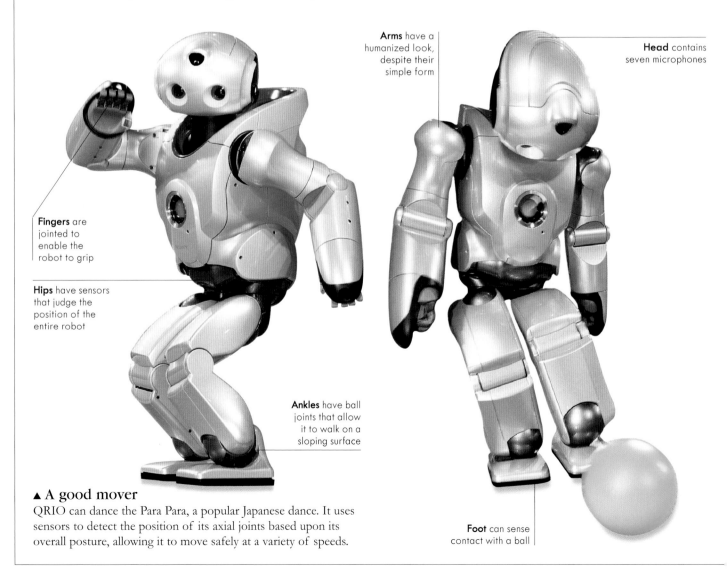

Arms have a humanized look, despite their simple form

Head contains seven microphones

Fingers are jointed to enable the robot to grip

Hips have sensors that judge the position of the entire robot

Ankles have ball joints that allow it to walk on a sloping surface

Foot can sense contact with a ball

▲ A good mover

QRIO can dance the Para Para, a popular Japanese dance. It uses sensors to detect the position of its axial joints based upon its overall posture, allowing it to move safely at a variety of speeds.

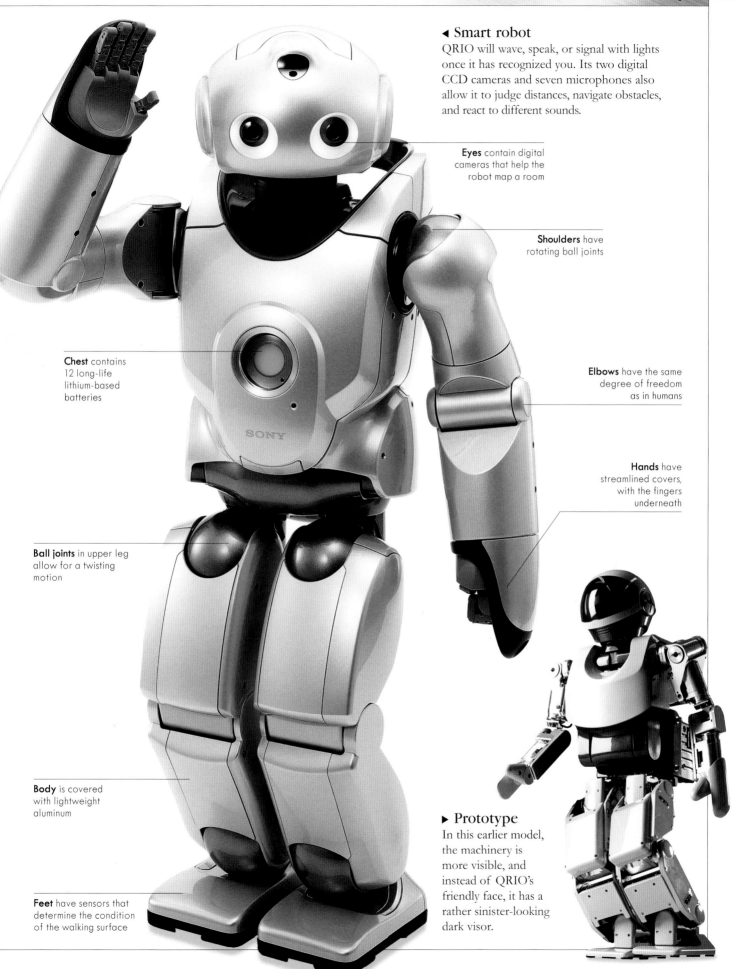

◄ **Smart robot**
QRIO will wave, speak, or signal with lights once it has recognized you. Its two digital CCD cameras and seven microphones also allow it to judge distances, navigate obstacles, and react to different sounds.

Eyes contain digital cameras that help the robot map a room

Shoulders have rotating ball joints

Chest contains 12 long-life lithium-based batteries

Elbows have the same degree of freedom as in humans

Hands have streamlined covers, with the fingers underneath

Ball joints in upper leg allow for a twisting motion

Body is covered with lightweight aluminum

▶ **Prototype**
In this earlier model, the machinery is more visible, and instead of QRIO's friendly face, it has a rather sinister-looking dark visor.

Feet have sensors that determine the condition of the walking surface

Nuvo

The ultimate executive desk toy, Nuvo is a robot designed for the switched-on mobile age. Call this functional humanoid from anywhere in the world and see what it sees—its scanner can be programmed to take photographs and then send them to your cell phone.

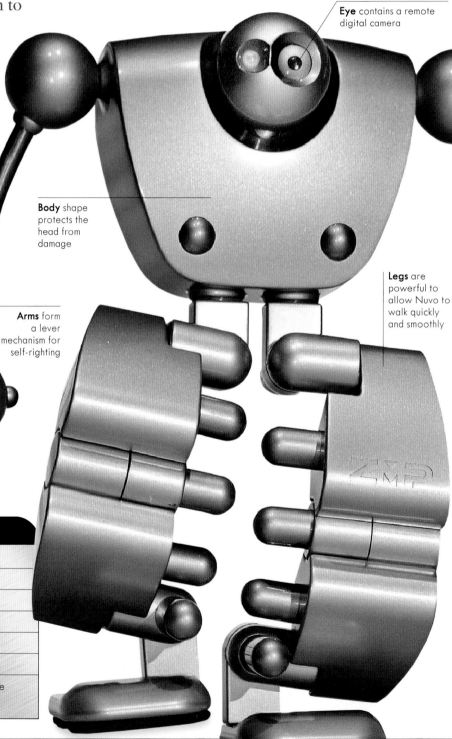

◄ Striking gadget
Nuvo is available in colors such as silver and red, and comes covered with a pattern of written messages. The soles of its feet have rubber tracks similar to those found on running shoes.

Eye contains a remote digital camera

Functional humanoid

Nuvo's distinctive body shape is an ingenious system of levers: it bends at the leg joints rather than the waist, walks fluidly, and can also pick itself up from a prone position. But Nuvo has brains as well as brawn, recognizing 1,000 voice commands, including "dance" and "bow."

Body shape protects the head from damage

Legs are powerful to allow Nuvo to walk quickly and smoothly

Arms form a lever mechanism for self-righting

► Body language
Nuvo is at the forefront of the new generation of beautiful, functional robotics. It looks like a fighter with his hulking frame and peering eye, yet it has some of the elegance of a designer accessory. Its ball-ended arms allow it to self-right, and the spherical motif is repeated in the head and eyes.

Specification: Nuvo

First manufactured:	2004
Country of origin:	Japan
Manufacturer:	ZMP
Height:	15 in (39 cm)
Power source:	Battery-operated
Intelligence:	Onboard CPU
Capabilities:	Walks, stands up, dances, obeys voice commands, records and sends digital images

Partner Robots

Toyota is the latest Japanese automotive manufacturer to see the future of technology as being humanoid in shape. Its new range of robots walks or rolls into the robotic companions market with a mission to care for the elderly and disabled, or for use in manufacturing, where it can be programmed to carry out a range of tasks.

Music-makers

These robotic partners are a giant leap ahead of many rivals—they are not just dexterous helpers, but also skilled musicians that blur the line between industrial and entertainment robotics. Their Linux operating system represents the collective work of thousands of software developers around the world.

◄ ▼ Right balance
The balance of both walking and rolling versions of the Partner is achieved by sensor technology derived from Toyota's automotive research.

Walking humanoid

Rolling humanoid

Hands are a simple sphere shape

► Adroit droid
This robot's lips mimic the movements of human lips. The dexterous fingers allow the robot to play a trumpet with great accuracy—no mean feat.

Specification: Walking Humanoid

First manufactured:	2004
Country of origin:	Japan
Manufacturer:	Toyota
Height:	4 ft (1.2 m)
Power source:	Lithium ion batteries
Intelligence:	Pentium III CPU, real-time Linux OS
Capabilities:	Walks, plays trumpet, obeys voice commands, uses hands to perform a variety of tasks

Joints have intelligent motors for flexible movement

Domestic Robots

Everyone is familiar with the futuristic concept of robots helping out in the home. Back in the real world, a new generation of "help" is revolutionizing twenty-first-century homes by working as cleaners, butlers, guards, and even gardeners. Rechargeable batteries, miniature sensors and cameras, and advanced programming and design make this dream a reality.

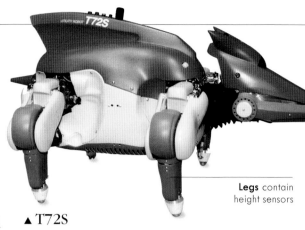

Legs contain height sensors

▲ **T72S**
The first production Banryu looks like a futuristic dinosaur, can be programmed to behave in guard or pet mode, and moves up to 50 ft (15 m) per minute.

Banryu guard robots

These guard dogs or, more accurately, dragons, are the result of collaboration between Sanyo and Tmsuk. The utility robots use voice-recognition technology and can be activated via their owners' cell phones. Banryus can roam around a home and will raise the alarm if they detect an intruder or the smell of smoke or gas. They can also be programmed to act as family pets, and will obey voice commands such as "sit" or "paw."

▼ **T7S Type 2**
This prototype can work for four-and-half hours on its batteries. It talks to its owner via a mobile phone, and can even beam pictures of any intruders it detects via a CCD camera.

Specification: Banryu T72S

First manufactured: 2002	
Country of origin: Japan	
Manufacturer: Sanyo/Tmsuk Co. Ltd.	
Height: 27 in (70 cm)	
Power source: Battery-operated	
Intelligence: Onboard microprocessors	
Capabilities: Walks, detects intruders, senses smoke and gas, is activated via cell phone	

Head is equipped with a CCD camera

Specification: Banryu T7S Type 2

First manufactured: 2002	
Country of origin: Japan	
Manufacturer: Sanyo/Tmsuk Co. Ltd.	
Height: 27 in (70 cm)	
Power source: Battery-operated	
Intelligence: Onboard microprocessors	
Capabilities: Walks, detects intruders, senses smoke and gas, is activated via cell phone	

Wheels help the robot to move forward smoothly

Legs can walk on uneven surfaces at up to 80 ft (25 m) per minute

RL 1000 mower

A simple but eco-friendly robot that mows grass without supervision, RL 1000 works only inside a wired perimeter. It can cut grass at six different mowing heights.

▶ Two-in-one
This mower not only cuts grass but mulches it as well, with zero emissions. Its mulching blades distribute the cuttings as the robot mows.

▶ Rechargeable robot
RL 1000 recharges at its docking station. It can then run for two hours.

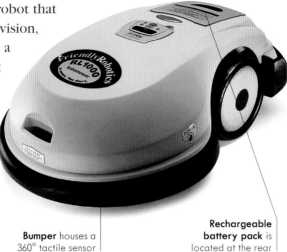

Bumper houses a 360° tactile sensor

Rechargeable battery pack is located at the rear

Specification: RL1000

First manufactured:	2003
Country of origin:	United States
Manufacturer:	Friendly Robotics
Height:	10 in (25 cm)
Power source:	Battery-operated
Intelligence:	Perimeter wire recognition
Capabilities:	Cuts and mulches grass within a wired perimeter

Cye Robot

This personal robot makes a useful butler. It can carry dishes, deliver letters, and conduct guests around the house. A special attachment even converts it into a vacuum cleaner.

▶ Mapped path
Cye Robot makes its way around a room using calibrated wheels equipped with wheel encoders. It generates its own map internally after navigating an area.

Entry point for vacuuming attachment

Wheels are serrated for better traction

Fender houses differential drive for independent wheel action

Specification: Cye Robot

First manufactured:	2003
Country of origin:	United States
Manufacturer:	Probotics Inc.
Height:	10 in (25 cm)
Power source:	Battery-operated
Intelligence:	Map-n-Zap software
Capabilities:	Carries objects, navigates, vacuum-cleans

DC06

This prototype Dyson vacuum cleaner can navigate a house with ease and can even remember where it has cleaned. It is safe around children and pets, since the sensory devices will pause it if they get too close.

▶ Clever cleaner
The DC06 can "think" for itself. It turns corners and has a failsafe system to prevent it from falling down stairs.

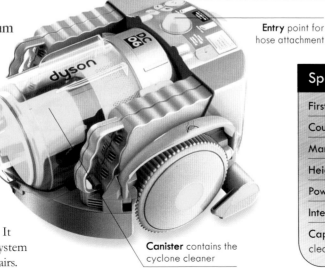

Housing contains three computers and 70 sensory devices

Entry point for hose attachment

Canister contains the cyclone cleaner

Specification: DC06

First manufactured:	2004 (prototype)
Country of origin:	UK
Manufacturer:	Dyson
Height:	10 in (25 cm)
Power source:	Battery-operated
Intelligence:	Three onboard computers
Capabilities:	Navigates autonomously, cleans by cyclone action

Robots at work

WORKING ROBOTS have come a long way since 1954, when George Devol designed the first programmable robot. The world is now teeming with them: there are robot golf carts, vacuum cleaners, lawnmowers, and even robot traffic cones. Another breed of working robots performs important labor-intensive or dangerous tasks.

▲ Programmed to work
Industrial robots perform endlessly repetitive tasks in automotive assembly, such as welding, spray-painting, and wheel-mounting.

Robots are an indispensable part of the modern workforce in a variety of industries. They are able to perform tasks that are too arduous or even simply too tedious for a human worker, and we can send them into hostile environments— from the ocean floor, to the crater of a volcano, to outer space—where human beings may not go.

In manufacturing industries, as well as being able to complete routine tasks efficiently, such as spraying paint, robots can now also perform micro-assembly tasks, packaging, labeling, placement, and wrapping.

▲ Sewer Access Module
This Swiss-made Ka-Te Systems sewer scanner has a camera, to relay pictures back to base, and a retractable arm.

Industrial robots are also used to handle and move toxic chemicals, volatile plastics, and hot metals.

Surveillance and research
The range of robot capabilities is being extended all the time. Robots are being used in military campaigns, for surveillance, to collect scientific data in hostile environments such as the Arctic, and for space station repair. To do this complex work, they need to be equipped with the latest cameras and sensors, and must have the ability to transmit a mass of data back to a central processor.

Built-in brains
Most early working robots were remote-controlled, and many still are, but the latest robots have an onboard computer (also known as B.O.B, or Brains On Board). Because

these robots have a memory, they can "learn" and operate without constant human intervention.

Until recently, most industrial robots were deployed in fixed locations such as factories or warehouses, performing a single, repetitive function. Now robots can walk, roll, crawl, or even fly, as well as navigate autonomously, so it is possible to program them to carry out more complex tasks.

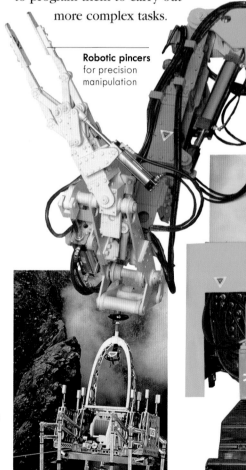

Robotic pincers
for precision manipulation

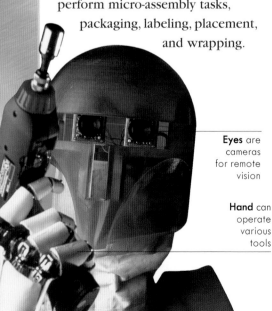

Eyes are cameras for remote vision

Hand can operate various tools

◄ Robonaut
NASA's Robonaut is designed to assist astronauts in external space station repair with its precision robotic hand.

► Dante II
This volcano explorer robot secured valuable readings from the pit of the Mount Spurr volcano in Alaska via a satellite link.

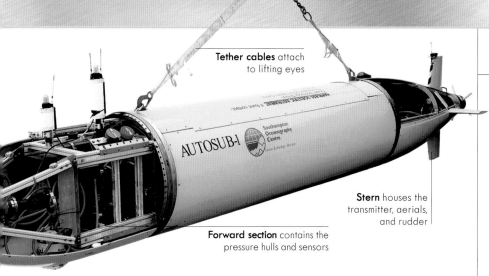

Tether cables attach
to lifting eyes

AUTOSUB-1

Southampton
Oceanography
Centre

Stern houses the
transmitter, aerials,
and rudder

Forward section contains the
pressure hulls and sensors

▲ Autosub-1 AUV
This robot is an autonomous robot submarine
used for scientific missions. It is nearly 23 ft
(7 m) long and can dive to 1,640 ft (500 m).

▼ T-52 ENRYU
This giant robot or "rescue dragon" can help in
disasters such as earthquakes, and can also act
as a firefighters' aide by cutting through debris.

Although robots are replacing us in
ever wider fields, there are certain
human actions that they are
currently unable to replicate and, as
yet, there are no "general worker"
robots that can switch from one
type of job to another. What is a
simple, intuitive task for a person
can be impossible for even the most
sophisticated robot. A typical
example is washing windows:
there are too many variables

tmsuk

Hydraulic
arm packs a
massive punch

Central
cabin is
for human
operator

for a robot to cope with,
such as different sizes, shapes, and
obstacles. However, a difficult and
time-consuming task, such as
precision welding of a car, is
straightforward for a robot, which
also has the advantage that it will
never get bored. Today, robots are
playing an increasing role in the
working lives of human beings, but
the day when they will actually replace
us is still very far in the future.

Surgical robots

The robotic surgical system da Vinci
was developed at Yale University
in Connecticut. Its telesurgery
devices are controlled by a
surgeon, sometimes remotely, via
a computer. The robot mirrors and
scales down the surgeon's actions
to reduce invasive procedures.

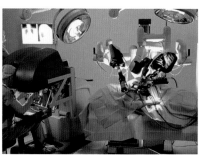

▲ Reaching the inaccessible
Here, a surgeon (far left) operates the
da Vinci system as a robotic extension of
his hands. It allows him to perform delicate
operations in inaccessible areas of the body.

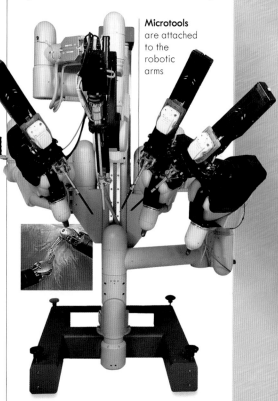

Microtools
are attached
to the
robotic
arms

▲ Extra hands
The five-armed da Vinci robot can use
microtools during an operation. The inset
picture shows the tiny robotic hands doing
incredibly precise work.

Security Robots

Robot security guards appeared over 50 years ago for three specific purposes: nuclear power systems surveillance and repair, warehouse and home surveillance, and assistance in hostile situations. Over time, advances in computers, miniaturization, and communications have increased their capabilities considerably.

OFRO

A companion to MOSRO, the OFRO security robot is designed for outdoor patrolling for periods of up to 12 hours at a time. It is used to monitor airport perimeters, military establishments, industrial plants, and other large areas where there are potential security threats. The top of the robot is full of sensors and is able to rotate through 360 degrees. Since it is mainly deployed in outdoor environments, OFRO is shielded from the effects of the elements by an all-weather protective casing.

MOSRO

Designed to patrol indoor areas such as malls and factories, this robot has a camera and a microphone, and can be equipped with up to 240 sensors that detect motion, gas, and smoke. It has an integral fingerprint scanner, and can issue verbal warnings in over 20 languages.

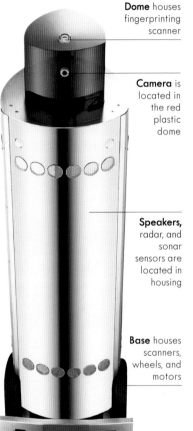

Dome houses fingerprinting scanner

Camera is located in the red plastic dome

Speakers, radar, and sonar sensors are located in housing

Base houses scanners, wheels, and motors

▶ **Simple exterior**
MOSRO's simple design masks complex electronics, and it can move at 2½ mph (94 km/h) for up to 18 hours between charges.

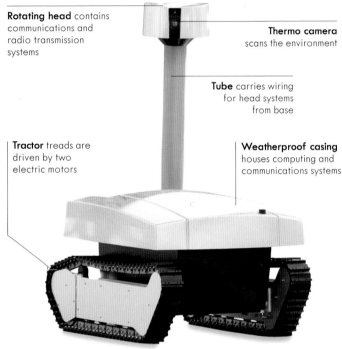

Rotating head contains communications and radio transmission systems

Thermo camera scans the environment

Tube carries wiring for head systems from base

Tractor treads are driven by two electric motors

Weatherproof casing houses computing and communications systems

▲ **Weatherproof watchman**
OFRO can withstand wind, rain, and snow as it goes about its appointed duties. Its caterpillar drive moves it quietly and safely through its patrol area, even on rough terrain.

Specification: MOSRO
First manufactured: 2003
Country of origin: Germany
Manufacturer: Robowatch Technologies
Height: 4 ft (1.2 m)
Power source: Battery-operated
Intelligence: Neural networks, Windows interface
Capabilities: Self-navigates, scans and takes fingerprints, issues warnings and communicates locally or via telecommunications

Specification: OFRO
First manufactured: 2003
Country of origin: Germany
Manufacturer: Robowatch Technologies
Height: 4 ft (1.2 m)
Power source: Battery-operated
Intelligence: Onboard DGPS receivers, WLAN/GSM transmission
Capabilities: Self-navigates, patrols, takes photographs, negotiates difficult terrain

Robart III

This formidable prototype robot has been developed by the US Navy as an advanced demonstration platform for experiments in nonlethal responses to intruders. It is controlled by a remote human operator. The main challenge facing this type of security device is how to determine the level of threat and then respond appropriately. Robart III is equipped with a camera and infrared sensors, and armed with an air-powered dart gun. It is the product of years of research and development, beginning in 1980 with Robart I.

Video camera is located on top of the head

◄ On target
A single remote operator can control Robart III's movements and weapon targeting simultaneously.

Polaroid tranducers decode sonar messages

Infrared sensor array devices are located in lower head

Casing holds wiring, computing devices, and communication links

Dart gun arm has a rotating shoulder and wrist

Compressed air is used for firing dart gun

Dart gun is pneumatic

▲ Rear view
The back shows the small caster-style wheel for swift direction changes, the pneumatics equipment for the gun, and the back of the rangefinder that is positioned on the arm opposite the dart gun.

Specification: Robart III

First manufactured:	1992
Country of origin:	United States
Manufacturer:	US Navy/ Bart Everett
Height:	3¼ ft (100 cm)
Power source:	Battery-operated
Intelligence:	Microcontroller
Capabilities:	Detects intruders, changes direction quickly, takes photographs, fires dart gun

Wheels are driven by electric motors

SICK

Millibots

A recent trend in robotic research is mini-robots that cooperate with each other and act as a team, maximizing their separate skills. Such robots could be deployed in sensitive locations, or in areas that are hard to access.

Tiny teammates

Carnegie Mellon University, through funding from DARPA, has initiated a project called "Millibots" to develop tiny robots that can carry out a multitude of functions as a team. These include taking pictures, carrying out surveillance, and taking sensor readings. Research to link the actions of the individual robot to the potential of team working is ongoing.

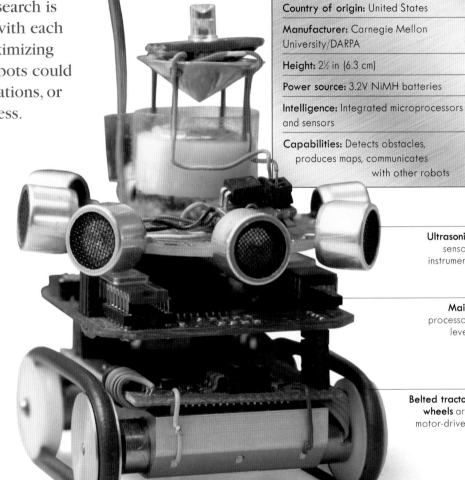

Specification: SonarBot	
First manufactured: 2001	
Country of origin: United States	
Manufacturer: Carnegie Mellon University/DARPA	
Height: 2½ in (6.3 cm)	
Power source: 3.2V NiMH batteries	
Intelligence: Integrated microprocessors and sensors	
Capabilities: Detects obstacles, produces maps, communicates with other robots	

Ultrasonic sensor instrument

Main processor level

Belted tractor wheels are motor-driven

▲ **SonarBot**
This robot has a sonar array to detect obstacles up to 20 in (50 cm) away. It helps to coordinate robot-to-robot ranging as the machines do their work.

▲ **Size matters**
Each specialized robot is 2–2¼ in (5–6 cm) long and is equipped with either a processor, sensors, or a camera.

▶ **A growing family**
As the Millibot project continues, new, more complex robots with increasingly specialized skills are being added to the team.

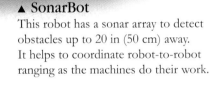

A robot community

The Millibots are being built with identical base structures and means of motion so that in time they will be able to repair each other. The aim is to build self-perpetuating robot communities that could be deployed at a distance (on another planet, for example) for up to a year. The next goal is to design a "mother ship" to receive and process feedback from the robots, and to supplement their functions. Research is continuing to develop both a sustainable source of energy and a recharging system.

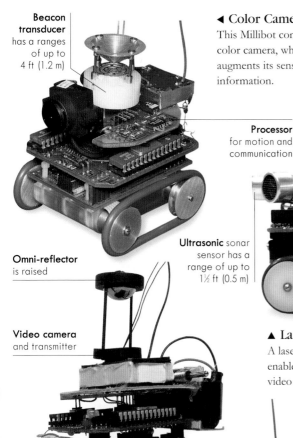

Beacon transducer has a ranges of up to 4 ft (1.2 m)

Omni-reflector is raised

Video camera and transmitter

◀ Color CameraBot
This Millibot contains a color camera, which augments its sensor information.

▶ BW CameraBot
A black and white camera and a video transmitter for off-board viewing are this robot's specialties.

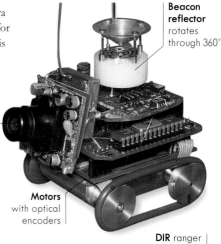

Beacon reflector rotates through 360°

Video camera is remotely enabled

Motors with optical encoders

Processor for motion and communication

Ultrasonic sonar sensor has a range of up to 1½ ft (0.5 m)

▲ LaserBot
A laser pointer and a spinning mirror enable this mini robot to perform video range-finding.

Mirror rotates for video range finding

DIR ranger covers 4 in–2½ ft (10 cm–0.8 m)

▲ OmniBot
The new OmniBot adds an omni-directional camera to the Millibot line, giving it increased versatility.

Beacon/sonar receiver array

Beacon/Sonar transmitter array

Processor is located at base

▲ DirrsBot
This Digital Infrared Ranging Sensor robot can follow walls and find openings. It has sonar elements for obstacle avoidance.

Pyro-detector scans for heat sources

Transducer turns readings into electrical signals

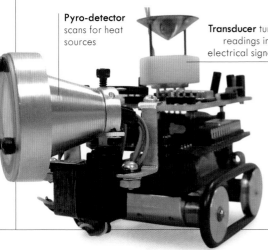

▲ Long Range SonarBot
This upgraded 8-element sonar array detects obstacles up to 3¼ ft (1 m) away much faster than previously.

◀ PyroBot
Used to locate fires, the Pyrobot houses a pyro detector that sweeps the area in front of it for heat sources.

Single module of the train

▶ Millibot Train
This train module helps the robots in difficult crossings of "large" obstacles. It is being developed in to a fully functional configurable unit.

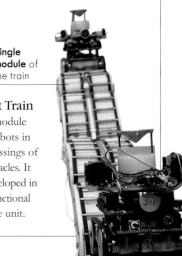

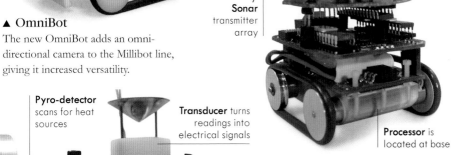

Space Probes

Autonomous robot devices that navigate the solar system exploring planets, moons, and comets have extended the frontiers of space research beyond the ordinary reach of human beings. Most of these projects involve a robotic probe that is sent to its destination aboard a spacecraft. The craft then relays the probe's data back to scientists on Earth.

Specification: Cassini-Huygens	
Date of launch: 1997	
Country of origin: United States/Europe	
Manufacturer: NASA/ESA	
Height of spacecraft: approx. 16 ft (5 m)	
Power source: Radioisotope thermoelectric generator	
Intelligence: Command and data, attitude and articulation subsystems	
Capabilities: Senses light and energy, senses magnetic fields, analyzes planet's surface, takes photographs	

Cassini-Huygens

The mission is to study Saturn and its rings, moons, and magnetic field. The spacecraft *Cassini* is heading toward Saturn's largest moon, Titan. When it gets close enough, it will eject a probe called *Huygens*, which, once it has landed, will relay photographs of Saturn and Titan to *Cassini*, which in turn will relay them to Earth.

▲ Perilous journey
Cassini's journey to Saturn takes it 746 million miles (1.2 billion km) through deep space. From there, it will launch *Huygens* toward Titan.

▶ Working partners
The spacecraft *Cassini*, from NASA, and the robotic probe *Huygens*, from the European Space Agency, are perfect partners. Their instruments are so sensitive that, unlike humans, they can detect magnetic fields and the tiniest dust particles.

Radioisotope thermoelectric generators provide power

Main engine provides forward thrust

Orientation thrusters allow the spacecraft to change direction

Antennae for two-way communication

Magnetometer is on 36-ft- (11-m-) long boom

▶ Ambitious mission
In this illustration, *Cassini* has just released the *Huygens* probe that is on its descent to Titan. The ring on the left of the spacecraft is the point of origin of the probe.

SMART 1 Moon probe

This orbiting laboratory probe searches the lunar landscape for frozen water and investigates the Moon's mineral and chemical composition. Launched by the *Ariane 5* rocket, it uses a revolutionary propulsion system in which solar electricity excites xenon, which then generates thrust by emitting a blue jet of ions. SMART stands for Small Mission for Advanced Research in Technology— missions designed to reduce the cost and complexity of space projects.

▼ Compact probe
SMART 1 is a relatively small craft, weighing just 807 lb (367 kg) and measuring 3¼ ft (1 m) on each side of the cube. Its ion propulsion system allows it to travel ten times as far as conventional rockets, thus reducing the size of the engine needed.

Specification: Smart 1

Date of launch:	2003
Country of origin:	Europe
Manufacturer:	European Space Agency/ Swedish Space Corp.
Height:	3¼ ft (1 m)
Power source:	Solar-electric propulsion, ion drive
Intelligence:	Spacecraft controller/8 cardas, 20 MHz microprocessor, 4 Gbit DRAM
Capabilities:	Conducts lunar analysis, tests for water

Solar panels are 50 ft (14 m) in length

Cubical vehicle has ion thruster on the underside

▶ Lunar readings
SMART 1's task is to take critical readings of the Moon by pointing its instruments toward the surface. It then relays them back to Earth.

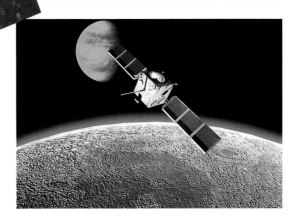

Rosetta Comet Probe

Rosetta will rendezvous with the comet 67P/ Churyumov Gerasimenko, situated between Mars and Jupiter, in the year 2014. The spacecraft is propelled by swing-by maneuvers drawing on Earth and Mars' gravitational forces. A lander named *Philae* will probe the physical properties of the comet as well as take photographs. *Rosetta*'s tryst with the comet will last for two years.

▼ Always in touch
As *Rosetta* orbits the comet, its instruments will always point toward the surface, and the giant antenna will always point toward Earth.

Specification: Rosetta

Date of launch:	2004
Country of origin:	Germany
Manufacturer:	ESA consortium
Height:	6½ ft (2 m)
Power source:	Solar array with 850-watt batteries
Intelligence:	Onboard computers
Capabilities:	Takes photographs, inspects comet with sensors (UV, infrared, radio, microwave) and spectrometers, runs on solar power

Antenna is more than 6½ ft (2 m) long

Legs are equipped with shock absorbers for a soft landing

▶ First contact
The probe will deposit its lander on the comet's surface to investigate it. This will be our first direct contact with a comet's surface.

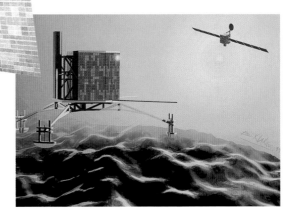

Robots on Mars

A new generation of robotic explorers is bringing us ever closer to Mars. Is there evidence of life in the planet's rugged terrain? Will we be able to colonize it in the future? Geological observation and mapping, and the probing of Martian soil, atmosphere, and weather by robots brings us closer to the answers.

Specification: Rover	
Date of launch: 2003	
Country of origin: United States	
Manufacturer: NASA/Jet Propulsion Laboratory	
Height: 4 ft 6 in (1.4 m)	
Power source: 140-watt solar panel	
Intelligence: Electronics module, inertial measure unit	
Capabilities: Rolls, recharges, takes photographs and X-rays, carries out geological analysis	

Mars Rovers

The Mars rovers act as robot geologists probing opposite sides of Mars on the twin NASA missions, *Spirit* and *Opportunity*. They use a panoramic camera to study the local terrain, a microscopic imager to obtain close-ups of the soil, and spectrometers to study minerals. A robotic arm uses a rock abrasion tool, while a magnetic array collects dust for analysis. The rovers have found evidence of the former presence of water.

▲ **Historical rock**
This rock, named Adirondacks after the mountain range in New York, was drilled to determine its chemical composition and geological origin. This was the Rover's first target after landing.

▶ **Navigating the unknown**
The *Spirit* rover scans the Martian surface with its robot arm extended and antennae and mast deployed. Below the outstretched solar array panels is the Warm Electronics Box (WEB) housing the computing system.

Mast holds four cameras

Low-gain antenna is for rough terrain

High-gain antenna is for precision targeting

Solar arrays are used for recharging energy

Beagle 2

Named after the ship that carried Charles Darwin on his ground-breaking voyage, the essential mission of this British-made lander was to carry out experiments on the surface of the planet in the hope of finding conclusive evidence of life on Mars. The Beagle had a large number of working tools in proportion to its light weight. Unfortunately, there has been no communication with the robot since its deployment in December 2003. An inquiry suggested too tight a build schedule and insufficient communication between teams.

Specification: Beagle 2
Date of launch: 2003
Country of origin: UK
Manufacturer: Consortium of UK universities
Height: 3 ft (90 cm)
Power source: 87-watt solar recharged batteries
Intelligence: Electronic computer module
Capabilities: Conducts biological studies using sensors, takes photographs, and compiles environmental data

▼ Thwarted mission
After landing, the Beagle 2 lander was to have studied the Martian atmosphere and looked for organic residues in the rocks and minerals.

▲ Artist's view
The Beagle 2 lander is seen here preparing to land on the surface of the red planet.

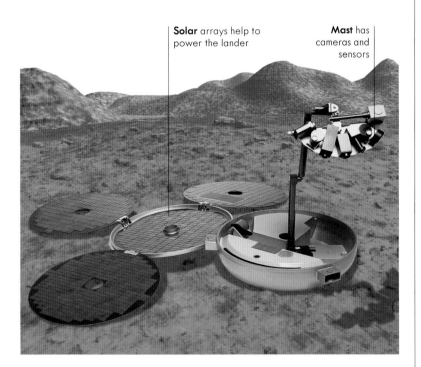

Solar arrays help to power the lander

Mast has cameras and sensors

Spider-Bots

NASA is conducting a series of experiments with the aim of developing miniature walking robots that can explore the surface of Mars. The idea is to make each mini-robot, or Spider-bot, a part of a team that reports to a base station. The prototype is able to navigate to a designated place according to instructions sent from Earth. It is hoped that advanced forms of Spider-bots will be able to interact with each other and carry out complex missions on the planet.

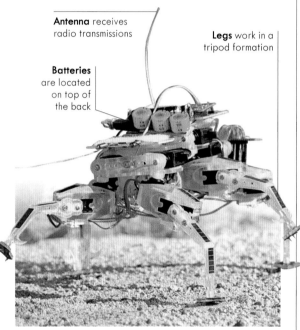

Antenna receives radio transmissions

Legs work in a tripod formation

Batteries are located on top of the back

▲ Techno spider
The Spider-bot is an advance on other types of mini-robots, since it walks rather than rolls across the surface, giving it greater ability to negotiate difficult terrain. It uses its six legs in a double tripod arrangement: three legs are placed on the ground for balance and three legs are in the air for motion. The robot has two micro-cameras that can survey its environment.

Specification: Spider-bots
First manufactured: 2002
Country of origin: United States
Manufacturer: NASA/Jet Propulsion Laboratory
Height: 6 in (15 cm)
Power source: Onboard batteries
Intelligence: Microprocessors
Capabilities: Walks, navigates the Martian surface, takes photographs

Future Robots

The challenges facing robotics engineers are primarily to create robots that are smaller, lighter, and more intelligent than ever before, and which have sustainable, efficient energy sources. The aim is to design robots that can travel to and carry out tasks in previously inaccessible places, from deep space explorers to microscopic machines that could travel within the human bloodstream.

8K ROM
processor operates drives and sensors

Circuit board
is the core of the vehicle

Front view   Actual size

Adkins Mini-Robot

A US-based project is developing a robot small enough to move through complex chemical apparatus and tooling. This requires innovative microelectronics and body design. The Adkins Mini-Robot, named after its creator, mechanical engineer Doug Adkins, is able to travel at slightly over 20 in (50 cm) per minute. It is completely untethered and so requires onboard intelligence. It weighs less than 1 oz (28 g).

Specification: Adkins Mini-Robot
First manufactured: 2001
Country of origin: United States
Manufacturer: Sandia National Laboratories
Height: ¾ in (15 mm)
Power source: Battery-operated
Intelligence: Multi-chip module
Capabilities: Moves in all directions, senses temperature

Specification: Entomopter
First manufactured: 2001
Country of origin: United States
Manufacturer: OAI
Wingspan: 5 ft (1.5 m)/4 ft (1.2 m)
Power source: Onboard fuel (refillable)
Intelligence: Microprocessor
Capabilities: Flies, takes photographs, communicates with refueling platform

Batteries
power the robot

Brass key
acts as a battery switch

Tractor
wheel system drives robot

Rear view   Actual size

▲ **Small is beautiful**
The stern of this "turn on a dime and park on a nickel" robot is dominated by its batteries. Its size has been reduced by eliminating the electronics packaging and using new techniques for body construction and wheel design. The body contains the batteries, two motors, axles, switches, and an electronic-embedded substrate.

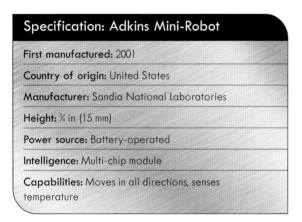

Entomopter aircraft and refueling platform

This flying robot is the work of Anthony Colozza and others at the Ohio Aerospace Institute and Georgia Institute of Technology. Currently at the concept stage, it is designed to make trips from a ground station and fly over the Mars landscape collecting low-resolution information over greater distances than a rolling or walking robot can. Its fuel system activates a motor drive that energizes its wings.

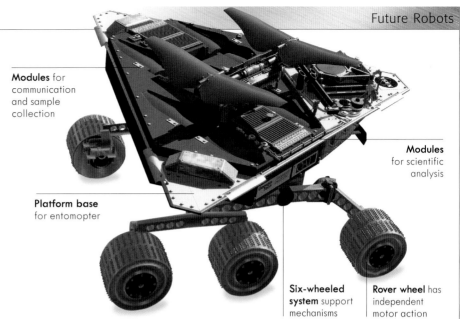

Modules for communication and sample collection

Platform base for entomopter

Modules for scientific analysis

Six-wheeled system support mechanisms

Rover wheel has independent motor action

▲ Teamwork

The Entomopter and its refueling platform work as a team, allowing the aircraft to carry only the fuel it needs for each mission. The platform guides the robot home once its work is complete.

Forward wing is part of energy-enhancing torsion system

Wings flap as the Entomopter prepares for landing

Rear wing is 180° out of phase with front wing to aid aerodynamics

Retractable antenna is in down mode

Legs cushion landing and add spring to liftoff

◀ Ready for landing

This artist's impression shows the Entomopter hovering over the refueling platform for a vertical landing after a low-resolution photography mission on Mars. The Martian landscape would not allow for a fixed-wing takeoff, so the vehicle would flap its wings about ten times per second in flight.

GALLERY: Robots: The New Generation

The latest robots are more sophisticated than ever before, mainly due to developments in artificial intelligence. The most advanced robots are autonomous, and can compute, navigate, and perform tasks without human interaction; others can be preprogrammed or operated by remote control. They range from bomb disposal robots to vacuum cleaners, carers, and even pets. Some have an industrial appearance, while others mimic humans or animals.

Electrolux Trilobite
This vacuum cleaner navigates a room autonomously, and when its power runs low, it returns to a recharging station.
- Date: 1997 ■ Country of origin: Sweden
- Height: 6 in (15 cm) ■ Manufacturer: Electrolux

Disrupter fires water into the bomb to disarm it

Video camera is used to aim the disrupter

HOBO
This Hazardous Ordnance Bomb Operator is a tele-operated bomb-disposal robot.
- Date: 1980s ■ Country of origin: Ireland
- Height: 2 ft 9 in (88 cm)
- Manufacturer: Kentree

Wheels are individually driven by separate motors

Tmsuk 04
A humanoid robot created for research purposes, Tmsuk 04 is remotely controlled via a PHS radio signal.
- Date: 1999 ■ Country of origin: Japan
- Height: 5 ft 3 in (1.6 m)
- Manufacturer: Tmsuk

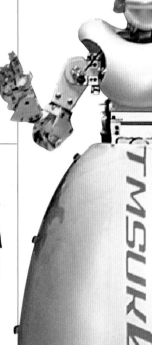

Robocleaner RC 300
This autonomous cleaner has dirt sensors and anti-fall sensors to help it avoid steps.
- Date: 2001 ■ Country of origin: Germany
- Height: 4¼ in (10.5 cm) ■ Manufacturer: Alfred Kärcher GmbH & Co.

Roomba
Run on intelligent navigation technology, this cleaner makes the transition to different kinds of flooring with ease.
- Date: 2002 ■ Country of origin: United States
- Height: 4 in (10 cm) ■ Manufacturer: iRobot

Face is able to express a variety of emotions

Kismet
This robot face is the result of research into human/robot interaction. It runs on 15 integrated computers.
- **Date:** 2001 ■ **Country of origin:** United States
- **Height:** 32 in (80 cm) ■ **Manufacturer:** MIT

Packbot
This remotely controlled military robot can be used in the field for rugged surveillance and bomb-disposal work.

- **Date:** 2000 ■ **Country of origin:** United States ■ **Height:** 7 in (18 cm)
- **Manufacturer:** iRobot/DARPA

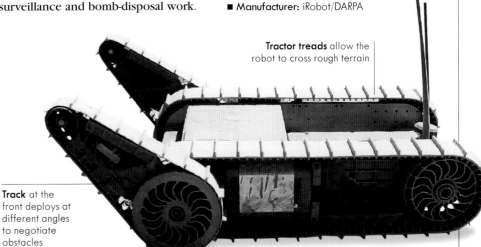

Tractor treads allow the robot to cross rough terrain

Track at the front deploys at different angles to negotiate obstacles

Wakamaru
Designed to take care of the elderly and disabled, this robot can talk, recognize faces, and access the Internet.
- **Date:** 2003 ■ **Country of origin:** Japan
- **Height:** 3 ft 3 in (1 m) ■ **Manufacturer:** Mitsubishi Heavy Industries

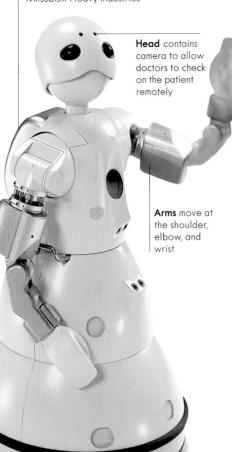

Head contains camera to allow doctors to check on the patient remotely

Arms move at the shoulder, elbow, and wrist

Sea Bass
Designed for a virtual aquarium, this computer-controlled fish is remarkably lifelike in appearance and movement.
- **Date:** 2003 ■ **Country of origin:** Japan
- **Height:** 3¾ in (12 cm) ■ **Manufacturer:** Mitsubishi

PartyBot
This personal robot can be preprogrammed to carry drinks, flash its eyes, and sweep the floor. It is controlled by a remote infrared unit.
- **Date:** 2003
- **Country of origin:** UK
- **Height:** 10 in (25 cm)
- **Manufacturer:** Firebox

Eyes contain flashing blue LEDs

Hexapod
This walking robot was designed by Matt Denton for a Harry Potter movie and will go into production in a modified form.
- **Date:** 2003
- **Country of origin:** UK ■ **Height:** 4½ in (11.5 cm)
- **Manufacturer:** Micromagic Systems

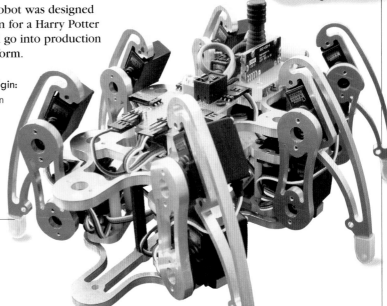

Legs each have a motor to control motion

Glossary

ACTUATOR Device for moving or controlling a tool, or part of a tool.

ANDROID Robot or automaton resembling a human.

ANTENNA Wire or other metallic structure that is used to receive or transmit signals, such as radio transmissions.

ARTIFICIAL INTELLIGENCE (A.I.) Computer programs that resemble, mimic, or take the place of, human thought processes. Also programs that allow devices to operate without human assistance.

AUTOMATION Technology that deals with the application of self-governing machines and systems.

AUTOMATON Doll or figure that is mechanically driven and controlled.

AUTONOMOUS Having self-power or self-control, or both.

BEAM Acronym for Biology, Engineering, Aesthetics, and Mechanics, a robotics control system that relies on simple electronics and mechanisms rather than microprocessors.

BIOMORPHIC Description of machines resembling the forms of living organisms.

BIONICS The science of extending biological principles to nonbiological systems.

BIT Within computerized systems, an abbreviation of "binary digit." Thus "8-bit" and "24-bit" refer to sequences of binary numbers that are eight and 24 characters long, respectively, within which data can be stored.

BOT Abbreviation of robot.

BUMP-AND-GO ACTION Mechanism within toys that causes a change of direction when the toy collides with an object.

CAD Acronym for Computer Aided Design, the use of computers to generate precise drawings, often using complex engineering calculations.

CAM Within engines, a rotating cylinder of irregular shape that moves other components.

CATERPILLAR TREADS Loops of textured grippers that fit over wheels to provide greater traction.

CLOCKWORK Accurate system of gears and cogs driven by the energy from a coiled

Meccano robot

spring, or swinging weights, as in a clock. In toys, usually the same as wind-up.

COMPARATOR Device for comparing things.

CONTROLLERS In robotics, computers that are used to control machines.

CPU Acronym for Central Processing Unit, the set of components within a computer within which data is processed.

CYBERNETICS The science of mechanical control systems and their similarity to the natural mechanisms found in living creatures.

CYBORG Being that combines human parts or functions with machine parts and functions.

DARPA The Defense Advanced Research Projects Agency, which makes defense robots in the United States.

DEGREES OF FREEDOM The number of directions allowed by a mechanical joint, defined by the number of rotational axes through which motion can be effected.

DIGITAL Description of how information is recorded and stored as sequences of binary numbers in computerized devices. More information can be stored in longer strings of numbers.

DIE-CAST An industrial, high-quality metal and plastic casting method.

DRIVE Within machines, a device for releasing stored energy and turning it into motion. Within computers, a device for storing electrical energy as digital data.

DROIDS Abbreviation of Android, and now a synonym for any type of robot. First used in the latter case in *Star Wars*.

DRONE Uncrewed robotic machine operated by remote control, or robot designed for mass manual labor.

ENDOSKELETON Internal skeleton or structure of a vertebrate creature or robot

FEEDBACK Looping a system's output back to its input, or source. Raw data is inputted and processed, and the modified data is returned to the original input.

GRIPPER Mechanism for holding an object that is distinct in its design from a humanoid hand.

GYROSCOPE Component containing a disc rotating on an axis, which allows it to balance either itself, or the larger device of which it is a part, regardless of external forces.

HUMANOID Anything resembling, or with the characteristics of, a human being.

HYDRAULIC Operated by means of water pressure.

INDUSTRIAL ROBOT Robot that is applied to manufacturing, such as welding, assembly, packaging, cutting, and painting.

INFRARED Form of electromagnetic radiation that is invisible to the human eye because of its long wavelength.

LED Acronym for Light Emitting Diode, a semiconductor device that emits light when an electrical current passes through it.

LINUX Free computer operating system named after programmer Linus Torvalds, who wrote the central "kernel" of computer code.

LITHOGRAPHY Printing process that was used to add color details to mass-produced tin toys inexpensively.

MEMORY STICK Small, portable digital data storage device used within some devices.

MICROCHIP Small wafer of silicon onto which are printed thousands of electrical components. A chip is the heart of a microprocessor, which is sometimes known as a chip. The more components that can be printed onto the silicon, the more powerful the processor will be. Advances in computer technology are rooted in this fact.

MICROPROCESSOR See Processor

MICROPHONE Device for capturing sound waves and turning them into electrical energy for amplification, or modification.

MIT The Massachusetts Institute of Technology, a major center of robotic research, digital media, engineering, and programming.

MORPHING Changing from one form or shape into another.

NANO Measurement of one billionth of something, often applied to the theoretical concept of microscopically small robots.

NASA Acronym for the National Aeronautics and Space Administration, responsible for the US space program.

OPEN SOURCE Computer code, such as the Linux operating system, that is not copyrighted by a organization or individual. Users are therefore able to modify the code, and to share the modifications freely, without infringing any company's copyright.

PNEUMATIC Powered, or filled, by pressurized air.

A TV Screen Robot

POSITRONIC Fictional term coined by Isaac Asimov, and now used widely in science fiction, to describe a particle-driven operating system within robots' brains.

PROCESSOR The component at the heart of a computer, containing a silicon chip, within which information is turned into millions of electrical instructions every second. Miniaturized versions are known as microprocessors.

PROGRAMMABLE Able to accept and follow sequences of preset actions.

PROTOTYPE An interim product built for evaluating and improving a product that is intended for wider release.

PROXIMITY SENSOR Device that measures the closeness of another object by bouncing signals off it and measuring the elapsed time before the signal is returned.

PULP FICTION Type of popular story intended for mass consumption in magazines or inexpensive books.

RADIO CONTROL Means of sending simple commands to a remote device, such as a toy, by transmitting radio signals.

REMOTE CONTROL Means of controlling a device from a distance, either by radio, or by signals sent directly to it through a wire.

ROBOT Machine that resembles a human being, or that is designed to carry out tasks in place of a human being. Derived from the Czech word *robota*, meaning "forced (or menial) labor." First used in 1920 in Karel Capek's play *R.U.R.*

ROBOTIC Pertaining to robots, or robotlike.

ROBOTICS The science of designing, building, or using robots.

SENSORS Devices that detect and/or measure changes in energy, such as heat, light, sound, pressure, radiation, and so on. For example, motion sensor, light sensor, etc.

SERVO Short for Servomechanism. Within a mechanical device, a component that minimizes the differences between the level of energy input and the level of energy output, ensuring smooth operation.

SIMULATION In computing, the act or process of representing an object or event in mathematical terms, and building a copy of it.

SMART MOTORS Motors with controllers that allow instantaneous stopping or starting, and incremental rotation. Also commonly known as intelligent motors.

SOFTWARE Set of codes designed to tell a computer how to process data that is designed for a specific application or task.

SOLAR ARRAYS Devices that capture and store heat and light from the sun and turn it into energy to power other devices.

Reproduction Smoking Robot

TELEOPERATOR Device that can be operated by a human, sometimes from a remote location, that copies the human being's precise actions.

TINPLATE Metal sheet material made from tin-coated steel.

TRANSDUCER Device that transforms energy from one form into another, such as electrical energy to mechanical energy.

TRANSISTOR Small electronic component that amplifies or switches electrical signals.

USB (PORT) Acronym for Universal Serial Bus, a computer interface that can receive and send data at high speed.

VOICE ACTIVATION System for turning sound waves at the frequency of a human voice into electrical signals.

VOICE RECOGNITION System that allows the robot to recognize characteristic speech patterns.

WIND-UP Description of a system of gears and cogs in which a key coils a metal spring. When the spring is released, the energy drives a mechanism, such as a toy's.

WIRELESS MODULE Communications module that is radio-based and needs no wiring. This capability is now being built into some microprocessors themselves.

Index

Fighting Robot

LEGO
Mindstorms Kit

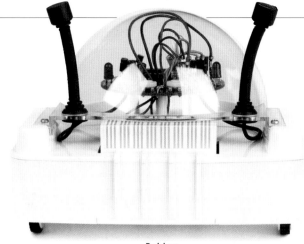

Rabbot

Acknowledgments

The author would like to thank the following:
I wish to thank Eric Alberga, as a toy expert and for pre-transformer toys; J. E. Alvarez, as a Transformer collector and expert on Transformer toys; Mark Bergin, as a toy robot collector and expert for sources. I am indebted to Robert Doornick, president and philosopher of International Robotics and an educational robot specialist and developer, as well as to Antoni Emchowicz for permission to photograph and use robots from his collection. I especially want to thank Patrick Karris, as a collector and as the Robby the Robot expert, for Robby information, pictures, and general knowledge of robot toys and film, and also the extremely cooperative curator, Joe Knedlhans, of the Toy Robot Museum, who is an expert on history of toy robots and robots generally. Many thanks to John Rigg, as the curator of The Robot Hut, for his knowledge of toys, films, and personal robot construction and as a robot expert and a great robot craftsman.

I also wish to thank the staff of DK for their persistence, patience, and very hard work, including editors Nicky Munro, Adèle Hayward, Neil Lockley, Corinne Asghar, and Chris Middleton, and designers Jenisa Patel, Karen Self, and Alison Shakleton; photographer Gary Ombler; as well as Carolyn Clerkin for her picture-gathering.

The publishers would like to thank the following:
Editorial: Corinne Asghar; May Corfield; Antonia Cunningham, Neil Lockley; Miranda Smith; Amber Tokeley
Design: Gillian Andrews; Mark Cavanagh; Michael Duffy; Jo Grey; Alison Shackleton; Dawn Young
Photographer: Gary Ombler
DTP: Sonia Charbonnier; Grahame Kitto
Proofreader: Alison Bolus; Indexer: Chris Bernstein
Loan of items: www.scificollection.co.uk at The Stamp Collectors Centre, 79 Strand, London WC2R 0DE
and Vincent C. Backeberg for use of his CAD image of Dewey
Thanks to the following for allowing us to photograph their collections: J. E. (Rikki) Alvarez; Village Doll and Toy shop; Antoni Emchowicz; International Robotics—Robert Doornick; Patrick Karris; The Toy Robot Museum—Jo Knedlhans; Robert Malone; The Robot Hut—John Rigg

Dalek

Picture Credits

1: NEC Corporation; **4:** NEC Corporation (cl); **5:** Eric Joyner (br); **6:** Advertising Archive (c); Lawrence Northey: www.robotart.net (bl); **7:** Eaglemoss Publishing Group/Simon Anning. (tl); Kawada Industries, Inc. (tr); SPL/Peter Menzel (br); **8:** The Art Archive/Topkapi Museum Istanbul/Dagli Orti (c); www.bridgeman.co.uk/Lauros/Giraudon (tl), (tr); **8-9:** Art Archive/Rijksmuseum voor Volkenkunde Leiden (Leyden)/Dagli Orti (b); **9:** Corbis/Stefano Bianchetti (c); RGA/ courtesy PAGU (br); **10:** Image courtesy of The L. Frank Baum Trust (cbl); British Film Institute (cbr); Paul Guinan (bc), (tr); MEPL (cal); **11:** Corbis/Bettmann (bcl), (crb); MEPL (cb), (tc); Science & Society Picture Library/Science Museum (bcr), (bl); **12:** Advertising Archive (cbl); Getty Images/Toshifumi Kitamura/AFP (tc); NASA (bc); Rex Features/Sipa Press (bl); Science & Society Picture Library/National Aeronautics & Space Administration (br); **13:** Carnegie Mellon University & Nasa (bcr); Reuters/Toshiyuki Aizawa (crb); SPL/NASA (tc), Peter Menzel (br); Sony Corporation (cb); SRI International (bcl); **14-15:** Corbis/Lucidio Studio Inc. (bk); **16:** Corbis/Bettmann (tl); **16-17:** Sony Corporation; **18:** Antoni Emchowicz/Paul Nunnely (c), (l), (r); **19:** Antoni Emchowicz/Paul Nunnely (l); **33:** Sotheby's Picture Library, London (r); **52:** DK Images/Tony Marshall (l); **52-55:** DK Images/TRANSFORMERS and all related characters are trademarks of Hasbro and are used with permission. © 2004 Hasbro. All rights reserved; **60:** Associated Press/Chiaki Tsukumo (c); PA Photos/EPA (bl), (crb); **61:** Associated Press/Tsugufumi Matsumoto (r); Topfoto.co.uk/UPPA (bl); **62-63:** Kitano Symbiotic Systems Project; **64-65:** Images courtesy of WowWee Ltd/Mark Tilden; **66:** PA Photos/EPA (bl); **66-67:** Sony Corporation; **67:** NewsCast Photo Library/Sony (cra); Rex Features/Tony Kyriacou (cr); Sony Corporation (br); **70:** Active Robots; (cr); iBOTZ: www.ibotz.com (tr); © The Lego Group. LEGO, the LEGO logo and the Minifigure is a trademark of the LEGO group, here used by special permission (cl), (tl); RobotsRus Ltd. (br); **70-71:** Corbis/Lucidio Studio Inc. (bk); **71:** Eaglemoss Publishing Group/Simon Anning. For full details of this robot project please visit the website on www.realrobots.com; **72:** iBOTZ (bc), (br), (tl); **73:** DK Images/Avanced Design, Inc. www.robix.com (c); Dr Robot, Inc. (br); iBOTZ (tr); JCM inVentures.com (tl); **79:** www.owirobot.com (br), (l); **80:** Dave Hrynkiw/Solarbotics Ltd. (tr); MCII Robot (bl); **80-81:** iBOTZ (b); **81:** Active Robots: www.active-robots.com (tl); Dave Hrynkiw/Solarbotics Ltd (tl); iBOTZ (br); JCM inVentures.com (cr); **82-83:** Eaglemoss Publishing Group/Simon Anning. For full details of this robot project please visit the website on www.realrobots.com (tr); **84-87:** © The Lego Group. LEGO, the LEGO logo and the Minifigure is a trademark of the LEGO group, here used by special permission; **88:** © Cube Co., Ltd, http://www.cam-system.jp (br), (cl), (tr); **89:** Dr Robot, Inc. (b), (l), (tr); **90:** Active Robots: www.active-robots.com (cr); JCM inVentures.com (cl); PA Photos/EPA (bc); **91:** Active Robots www.active-robots.com (cl), (tl); iBOTZ (cr), (tr); RobotsRus Ltd. www.RobotsRus.com (bc); **92:** Copyright © 2004 BattleBots Inc. All rights reserved. BattleBots is a registered trademark of BattleBots, Inc. Photographer: Daniel Longmire (cr); © 2001 Imagi. All rights reserved. (bl); Kobal/Walt Disney (cl); MEPL (tr); Christian Ristow (tl); RGA/courtesy Universal Pictures (bl); **92-93:** Corbis/Lucidio Studio Inc. (bk); **93:** Rex Features/Peter Brooker; **94:** Advertising Archive (bl); Kobal/Republic (tr); RGA/courtesy Columbia/Touchstone Pictures (br); **95:** Advertising Archive (tcl), (tl); Aquarius Library/Dreamorks (c); Getty Images/Toru Yamanaka/AFP (cr); RGA/courtesy De Laurentiis/Marvel Productions/Toei Co (br); **96:** Rex Features/SNAP (SYP) (ca); RGA/courtesy UFA (bl); **96-97:** Kobal/UFA; **97:** RGA/courtesy UFA (r); **98:** Kobal/20th Century Fox (cb); **99:** Kobal/20th Century Fox (bc); **100:** Rex Features/SNAP (SYP) (car), (cbl); **100-101:** Corbis/John Springer Collection; **101:** Kobal/MGM (br), (cr); RGA/courtesy MGM (tr); **102:** Kobal/AARU Prods/ Moviestore Collection (bl); Rex Features/Chris Balcombe (br); **103:** DK Images/British Film Institute/Museum Collection (l); **104:** British Film Institute; **105:** Kobal (bl), (br), (cr), 20th Century Fox/CBS Television (cl); RGA/New Line Cinema (tr); **106:** Advertising Archive (cl); Kobal/Universal (tr); RGA/courtesy Universal Pictures (cr); **107:** Kobal/Glen A Larson Prods/Universal TV (bl), (tr); **108-109:** Kobal/Lucasfilm/20th Century Fox (c); **109:** RGA/courtesy LucasFilm (tr); **110:** Kobal/Lucasfilm/20th Century Fox (cl); **111:** Aquarius Library/20th Century Fox (bl); DK Images/© 2004 Lucasfilm Ltd. & © 2004 Hasbro, Inc. All rights reserved. Used under authorization. (tr); Kobal/Lucasfilm (cla), (tcl), Lucasfilm/Keith Hamsher (crb), Lucasfilm/20th Century Fox (c), (cal), (ccr), (cl), (clb), (fcla), (fcr), (tl); **112:** Topfoto.co.uk (bl), UPPA Ltd (tr); **113:** Rex Features/London Weekend Television (br), (cl); **114:** BBC Archive (clb), (r); **115:** BBC Archive (bl), (cr); RGA/courtesy BBC (tr); **116:** Kobal/Carolco (bl); Rex Features/Masatoshi Okauchi (c); **116-117:** Rex Features/Peter Brooker; **117:** Aquarius Library/Warner Bros (br), (cra); RGA/courtesy Carolco Pictures (cr), (crb); **118:** RGA/courtesy Orion Pictures (bl), (cr); **118-119:** Kobal; **119:** Kobal/Orion (br), (cr), (tc), Orion/Deana Newcomb (tr); **120:** Corbis/Richard Blanshard (bl); Kobal/Danziger Productions (tr), Paramount (br), Rankin-Bass-Toho (cl), Rankin-Bass-Toho/Universal (cla); **121:** Kobal/Walt Disney (bl); RGA/courtesy 20th Century Fox Filmed Entertainment (cr), courtesy ITC (tc), courtesy MGM (br); **122:** Kobal/Paramount Television (br), (cr), (tr); RGA/courtesy Paramount Pictures (bc); **123:** Rex Features/Snap (SYP); **124:** Rex Features/20th Century Fox/Everett (bc), (bl), (ca), (cl), (clb); **125:** Rex Features/20th Century Fox/Everett (br), (l); **126-127:** © 2003 Create TV & Film Limited. LITTLE ROBOTS is a trademark of Create TV & film Limited. LITTLE ROBOTS is based on the book by Mike Brownlow, published by Ragged Bears Publishing; **127:** DK Images/with permission from Create TV & Film Ltd (tr); **128:** 4Kids Entertainment International Ltd (bl); Advertising Archive (bc); Kobal/Universal (cr), Universal TV (tr); RGA/courtesy Warner Bros (br); **129:** British Film Institute (bl); Kobal/Tri-Star (br), Walt Disney (cr); **130:** Atari (bc), (bl), (br), (ca), (cb), (tr); **131:** Metro3D Europe Ltd (bc), (bl), (br), (cb); Courtesy of TDK (ca), (cla), (cra), (tc); **132:** MEPL (bl), (cr); **133:** Advertising Archive (br), (c), (cr), (tl), (tr); MEPL (bc); **134-135:** © Michael Whelan; **136-137:** Images courtesy of Eric Joyner. For more info visit www.ericjoyner.com; **138-139:** Images courtesy of Karl Egenberger/Envision Design, Inc. (http://envisiondes.com/robotart); **140-141:** Photos by the artist, courtesy of www.claytonbailey.com; **142-143:** Christian Ristow: www.christianristow.com; **144-145:** Images courtesy of Lawrence Northey: www.robotart.net; **146:** Federation of International Robot-soccer Association (cla), (tr); Reuters/Max Rossi (bl); © The RoboCup Federation (cr); **147:** Federation of International Robot-soccer Association (cra), (tc); Reuters/Max Rossi (br), (cr); © The RoboCup Federation (bc); **148-149:** Copyright © 2004 BattleBots Inc. All rights reserved. BattleBots is a registered trademark of BattleBots, Inc. Photographer: Daniel Longmire; **150:** Advertising Archive (cl); Kobal/Renown (bc), Republic Pictures (c); **151:** Photo by the artist, courtesy of: www.claytonbailey.com (cr); © 200 Imagi. All rights reserved (bl); Kobal/Carolco (br), Lucasfilm/20th Century Fox (tl), Nu Image (cl); Image courtesy of Lawrence Northey (bc); David K. Rose: www.grfxmonkey.com (br); **152:** Aethon Inc. (br); Rex Features/Sipa Press (cl); RoboWatch Technologies (tr); SPL/James King-Holmes (bl), Peter Menzel (tr), Sam Ogden (cr); **152-153:** Corbis/Lucidio Studio Inc. (bk); **154:** ABB Ltd. (bl); Carnegie Mellon University (c); Intuitive Surgical, Inc. (tl); RoboWatch Technologies (bc); SPL/Colin Cuthbert (tr), James King-Holmes (br); **155:** Associated Press (br); Fujitsu Automation Ltd (cl); Reuters/Yuriko Nakao (tr); Waseda University/Atsuo Takanishi Lab (tl); **156:** NEC Corporation http://www.incx.nec.co.jp/robot (bl), (clb), (cr); **157:** Fujitsu Ltd (cr); Rex Features/Sipa Press (l); **158-159:** International Robotics Inc.; **160:** Honda (UK) (c), (cl), (cr); **161:** Getty Images/Koichi Kamoshida (br); Honda (UK) (c); **162:** Reuters/Kimimasa Mayama; **163:** Fujitsu Automation Ltd (bl), (c), (tl), (tr); **164-165:** Getty Images/Junko Kimura; **165:** Getty Images/Junko Kimura (tr); **166:** Getty Images/Yoshikazu Tsuno (bl), (br); **167:** Corbis/Tom Wagner (br); Sony Corporation (l); **168:** Getty Images/Toru Yamanaka/AFP (bc); Reuters/Toshiyuki Aizawa (tc), (tcr); **169:** Getty Images/Junko Kimura (tcr), Toru Yamanaka/AFP (br); Reuters/Eriko Sugita (cra); **170:** Associated Press/Tsugufumi Matsumoto (tr); Getty Images/Koichi Kamoshida (bc); **171:** Aethon Inc. (c); Courtesy of Friendly Robotics (tc), (tr); Rex Features (br); **172:** Carnegie Mellon University & Nasa (br); Corbis/Charles O'Rear (tr); Getty Images/Newsmakers (c); SPL/Peter Menzel (bl); **172-173:** Rex Features (b); **173:** DK Images/The Natural Environment Research Council who funded Autosub and Nick Millard of the Southampton Oceanography Centre, UK (tl); Intuitive Surgical, Inc. (tr); SPL/Peter Menzel (cra); **174:** Getty Images/Alexander Heimann (cr); RoboWatch Technologies (cl); **175:** Images courtesy of Space & Naval Warfare Systems Center, San Diego (cl), (r); **176:** Carnegie Mellon University/Mark Portland; **177:** Carnegie Mellon University/Bob Grabowski; **178:** NASA (bl); SPL/David Ducros (c), (cra); **179:** Rex Features (bl), (cl); SPL/Christian Darkin (cr), Erik Viktor (br); **180:** Rex Features/Sipa Press (bc); SPL/NASA/JPL/Cornell (cr). **181:** NASA (cr); SPL/European Space Agency (bl), (c); **182:** Douglas R. Adkins (cr), (tr); **182-183:** SPL/Rob Michelson/GTRI (b); SPL/Rob Michelson/GTRI (tr); **184:** DK Images/Kate Howey & Elgan Loane of Kentree Ltd, Ireland. (cl); Electrolux (tr); iRobot Corporation (bc); Kärcher UK Ltd. (bl); Rex Features/Tony Kyriacou (br); **185:** Associated Press (tr); Party Bot. Available from www.firebox.com at £39.95 (cr); Getty Images/AFP/Yoshikazu Tsuno (tc); Matt Denton: www.micromagic-sys.com (br); Mitsubishi Heavy Industries, Ltd. Japan. (bl); SPL/Sam Ogden (tl); **189:** © The Lego Group. LEGO, the LEGO logo and the Minifigure is a trademark of the LEGO group, here used by special permission. All other images © Dorling Kindersley. For further information see: www.dkimages.com

Meccano robot